Stop Global Warming
Change the World

Jonathan Neale

Also by Jonathan Neale

Memoirs of a Callous Picket
The Cutlass and the Lash
The Laughter of Heroes, a novel
Mutineers, a novel
The American War: Vietnam 1960-1975
Tigers of the Snow
Lost at Sea, a novel for children
You Are G8, We Are 6 Billion
Himalaya, a novel for children
What's Wrong with America?
Globalisation, Writing and Resistance (editor)

STOP GLOBAL WARMING

Change the World

Jonathan Neale

bookmarks publications

Stop Global Warming: Change the World—Jonathan Neale
First published in July 2008 by Bookmarks Publications
Copyright © Jonathan Neale

ISBN 9781905192373

Cover photographs taken by Soo Hyun Lim on Global Action Day
protests against global warming, December 2005, South Korea.

Cover design by rpm design
Typeset by Bookmarks Publications
Printed by Cambridge Printing

Contents

For
Linda Maher
and
Phil Thornhill

Introduction

WE can't completely stop climate change. But we can stop climate catastrophe—the swift feedback processes of "abrupt climate change". If we don't stop abrupt change, many species will be exterminated, and hundreds of millions of people will die of drought, hunger, thirst, disease, repression and war.

The main cause of global warming is carbon dioxide (CO_2) from burning oil, gas and coal. To stabilise the CO_2 in the air at safe levels we need to cut that burning by at least 80 percent per person in the richer countries within 30 years or less. That's a tall order, but it can be done. We have to cover the world with wind turbines and solar power. And we have to cut energy use. Most energy goes on buildings, transport and industry. The most important solutions are home insulation, turning off the air conditioning, replacing cars with buses and trains, and regulating industry.

George Bush and other world leaders say we can't make those changes. It will cost too much, they say. Americans will lose jobs, Bush says. We will all face a drastic fall in our standard of living, they say, and ordinary people will never stand for that. So, they argue, the politicians cannot act.

But stop for a moment. Think what "cost too much" means. It means real working people will be paid dollars, pounds and rupees for building wind turbines, insulating houses and laying railroad track. Cost too much means more jobs.

Look at what happened in the Second World War. Every major power redirected their whole economy to make weapons and kill as many people as possible to win the war. That meant millions more jobs, and it pulled the world out of the Great Depression. We need the same thing again, on a global scale, but this time to save as many lives as possible.

The money is there. The world spends a trillion dollars a year on arms and the military. There are enough people in the world who need the work. We don't need to sacrifice to stop global warming. Instead we need an assault on global poverty.

However, almost all the governments and corporations of the world have spent the last 30 years arguing for "neoliberalism" and "globalisation". These words boil down to some simple ideas. The first is, "Private good, public bad." The second is, "Profit matters more than human need." The third, and most important, idea is, "You don't have to like the market, but there's nothing you can do about it. There is no alternative."

The idea that you can't resist the market is the dominant establishment idea of our time. It is the strongest weapon the rich and powerful now have. They will not give it up easily. But if governments intervene for the climate on a global scale, people all over the world will say, "If we can do that for the air, why not for the hospitals? For the schools? For my pension?"

The rich and powerful don't want people thinking like that.

Some corporations also have special reasons to resist any effective action on global warming. In 2007 Wal-Mart was the largest corporation in the world, followed by Exxon Mobil (2), Shell (3), British Petroleum (4), General Motors (5), Toyota (6), Chevron (7), DaimlerChrysler (8), Conoco Phillips (9), and Total in tenth place.[1] That's six oil companies, three car companies, and a chain of suburban stores with vast parking lots. It's also a formidable array of corporate power. Action over climate change means their corporate death.

Bush, Cheney and Rice represent these carbon corporations. They are doing everything they can to prevent action on climate change. However, many other rich and powerful people now want action. They own the world and don't want to ruin it. But they can't bring themselves to defy the market. So the measures they advocate are nowhere near enough to solve the problem.

The Kyoto treaty, for instance, calls for 5 percent cuts when we need at least 60 percent globally, and even that 5 percent is unenforceable. Or look at Al Gore's film *An Inconvenient Truth*. The first 90 minutes are a magnificent and terrible warning to the world. The last minute is about what we can do—a list of tiny actions that won't make a difference.

If we don't act, though, the power of the market and the corporations will turn climate disasters into human catastrophes. Global warming will mean heat waves, superstorms, floods and droughts. In our existing global society when the crops fail in a poor country people will starve. In our global society refugees come up against borders policed by men and women with machine guns. The tents of the refugee camps stretch for miles and for years. On the other

Stop Global Warming: Change the World

side of the border, racism increases to justify keeping out the needy. In our society global warming will mean war. Change the balance of geographical power, and the great powers and the small will go to war to restore that balance. We are living through the wars for oil. We will live through the wars for water.

The recent climate disasters in New Orleans, Darfur, Bangladesh, and many other places are only a taste of the future.[2] The rich and powerful will effectively try to make the rest of us pay the price for climate change. Poor people will kill each other for what little is left, and human decency will drown in the rising waters.

In short, we have the technology to solve the problem, but the corporations and the powerful can't or won't act. So we have to mobilise the only force that can challenge them, the six billion people of the world. Until now environmentalists have mainly lobbied governments and educated the public. We now need a mass movement to force the politicians to act, or to replace them with people who will. That movement has begun. It is still small, but it is found on every continent, and is growing fast. This book is part of that movement.

THE SCALE OF
THE PROBLEM

CHAPTER I

Abrupt Climate Change

THIS chapter explains the threat of abrupt climate change. This is a book about the politics of climate change, but to understand the politics we have to start with the science. We need to establish the scale of the problem we face. That scale determines the solutions. I do not argue for broad and serious measures in this book because I prefer them. I do it because the threat of abrupt change—climate catastrophe—means we must have drastic solutions.

If you already understand the science of abrupt climate change, you can skip or skim this chapter and go on to the next.

Why abrupt change matters

On first hearing, abrupt climate change sounds like science fiction, or some fantasy environmentalists have invented to terrify people into action. So people often discount what they hear—inside they are suspicious. That's why it's important to understand that scientists do not fear abrupt change because they have been running computer scenarios and speculating about the future. They fear it because they now know that global warming has happened very fast many times in the past. Often most of this warming took place in 20 years or less. Sometimes the majority of the warming happened in three years. The indications are that we are getting close to a point where this will happen again.

Abrupt climate change, like slow climate change, will mean rising ocean waters, rising temperatures and rapidly changing ecologies. But during the period of abrupt change the climate will become much more unstable. Extreme climate events—storms, floods, heat waves and droughts—will become more common and more intense. With abrupt climate change we will see many great hurricanes, many floods and many droughts in the same year—and in the next year, and the one after that.

Think of the effects of Hurricane Katrina in Louisiana. Or the cyclone that killed 300,000 people in Bangladesh in 1970. Or the drought that has affected the African Sahel, from Ethiopia to Mali, over the last 40 years. Now think of dozens of major cyclones over three years, some of them stronger than any we have ever seen. Think of rising sea levels flooding not just New York, London, Shanghai, Amsterdam and Mumbai, but dozens of cities, deltas and coastal plains. Combine that with the failure of the South Asia monsoon, and urban heatwaves and forest fires over much of the world.

It is impossible to guess how many would die—probably hundreds of millions. Greater numbers will be degraded and brutalised by what they must watch and live through in order to survive. These will not be simply natural disasters. Governments are likely to react by using force to make the poor and ordinary workers pay the very large price of destruction. People may resist collectively. But it is just as likely that they will fight each other for the scraps to survive. Cruelty often follows in the wake of want.

The key thing here is *timing*. People, and societies, can change and adjust to deal with disaster. But to do that they have to learn collectively, and they have to argue and contest the political solutions to new problems. With global warming, there will not be that time. And the rich and powerful will panic.

This will not be Armageddon. Human life will recover, and go on. Many other species will not be so lucky. Plants, trees, animals and fish adapt to changing temperatures by moving. Some move up in altitude—these will die off as they reach the mountain tops. Those in the far north and far south will likewise have nowhere to go. Animals can move only short distances each year, and plants and trees even less. If the temperature rises too quickly, they will be stranded and extinct.

Of course species have survived periods of abrupt climate change in the past. What will make this one different is that human settlements block the potential routes for migration. Moreover, all species survive in niches as parts of a complex ecological system. If some species are killed off, and the escape routes of others are barred, then the ones who do make it through to the north or into the hills will find themselves isolated from their old food sources.[1]

Carbon dioxide

That's why abrupt climate change matters. To explain why scientists are afraid of it, we have to start with the basic science of global warming.[2]

Global warming is caused by two main "greenhouse gases"—carbon dioxide and methane. Carbon dioxide is more important. It is made from one molecule of carbon (C) joined to two of oxygen (O)—CO_2. As far back as we can go in the history of the planet, the more CO_2 there was in the air, the warmer the temperature. This is because CO_2 allows radiation to pass down from the sun, but stops some of the radiation from the earth rising back into space. That radiation is trapped as heat, and the earth warms.

CO_2 is not common in the atmosphere. At the moment, even with global warming, carbon dioxide makes up just 380 molecules out of every 1,000,000 in the air (or "380 parts per million" as the scientists put it). But that small proportion makes a large difference.

For hundreds of thousands of years the earth has gone back and forth between ice ages and warm periods. During the ice ages there were about 180 parts per million of CO_2 in the air. In the warm periods there were 280 parts per million.

Two hundred years ago the earth was in a typical warm period. Then the industrial revolution began. Humans began burning more and more coal, then oil and then natural gas. Coal is mostly carbon. Oil and gas are made from carbon and hydrogen. When coal, oil and gas burn, the carbon (C) joins with oxygen (O) from the air and makes CO_2. The CO_2 created by this process goes into the air, and is called carbon dioxide "emissions".

Some of these CO_2 emissions are soaked up by two natural escape routes. One route is that trees and plants take CO_2 out of the air. They use it to make carbohydrates—"carbs", the basic stuff of plants. So the more CO_2 in the air, the more plants and trees. The other natural escape route is via the oceans. CO_2 passes easily from the air to the water. Small organisms in the ocean use carbon to grow and to make their shells. When these organisms die, some of their bodies and shells drift to the floor of the ocean and are trapped there. These escape routes on land and at sea are called "carbon sinks".

At the moment human action puts about 3.5 parts per million of CO_2 into the air each year. The two carbon sinks absorb about 1.4 parts per million. About 2.1 parts per million remain in the air. But carbon dioxide is a stable gas, and does not break down easily. So the extra 2.1 parts per million we put up there each year stay in the atmosphere for between 100 and 200 years.

In the 200 years since we started seriously burning coal, oil and gas, the amount of CO_2 in the atmosphere has gone from 280 to 385 parts per million. It's the same level of increase that the earth saw in the move from ice ages to warm periods.

Methane

Methane is the second most important greenhouse gas. It is one carbon molecule and four hydrogen molecules—CH_4. It is much rarer than CO_2. There is over 200 times more CO_2 in the atmosphere than there is methane. One reason is that methane is an unstable gas. It breaks down in contact with ozone. So methane only lasts an average of about 12 years in the air, compared to 100 to 200 years for CO_2.

The problem is that methane has a much stronger warming effect than CO_2. A molecule of methane, over its whole lifetime, has about 20 times the warming effect of CO_2. But in the first ten years the molecules are up there, methane has 100 times the effect of CO_2.[3]

Over the lifetimes of carbon dioxide and methane, CO_2 has a much bigger effect, because it is up there longest. This means that in the long run carbon dioxide is the main problem. It now accounts for 70 percent of total man made warming, and methane for about 13 percent. But in the short run, cuts in methane emissions would make a large immediate difference. And methane is a particular worry for abrupt warming, because feedback effects that release large amounts of methane could make a very large, very fast difference.

The amount of methane in the air has doubled since 1800. The good news is that methane emissions are now falling slightly.[4]

Methane emissions come from two main sources. Natural gas is almost all methane, and it leaks from coal mines, oil fields, gas fields, gas pipelines and power stations. The other main source of methane is biological decay. Plants, trees and animals are built from carbs. When they decay in contact with the air, that carbon joins oxygen to make CO_2. But when life decays in places where there is no air, the carbon (C) combines with hydrogen (H) to make methane (CH_4). This happens with organic waste in landfills. It also happens in swamps, under lakes and in flooded rice paddies. And it happens in the stomachs of animals that digest their food slowly, and especially in cattle that chew the cud. Mercifully, almost all these sources of methane emissions can be cut dramatically and relatively easily. I explain how in Chapter Ten.

Other greenhouse gases

Carbon dioxide and methane are the main greenhouse gases, but there are many others. Of these, the most important is nitrous oxide. This comes mainly from the use of fertilizers, car exhausts

The Scale of the Problem

and industrial processes. As with methane, cuts in emissions of nitrous oxide and the remaining greenhouse gases are also relatively easy.

Abrupt change in the past

To repeat, the reason scientists are worried is not speculation about the future, but discoveries about what happened in the past.

Traditionally scientists assumed that climate change happened slowly. Then two teams of scientists, one European and one American, began drilling down into the Greenland ice sheet in 1989. What they found changed everything.

Greenland is the best place in the world to discover how the climate changed year by year in the past. In Greenland winter and summer snow look different and have different chemical properties. They harden into ice that shows the same alternation between winter and summer forms. This meant that as the scientists drilled down they could read the cross-sections of ice year by year, the same way we can read tree rings. The scientists also ran chemical tests on the ice and the air bubbles trapped inside it. Those tests told them what the temperature was year by year, and how much carbon dioxide, methane and water vapour there was in the air. The Greenland ice sheet is two miles and 110,000 years deep. The scientists were expecting gradual change. What they found were 24 periods of major temperature change. When the earth was cooling, the change was often gradual. When the earth warmed, most of the change often happened within 20 years or less. 10,660 years ago, at the end of the last ice age, the majority of the warming in Greenland happened in just three years.[5] That's very abrupt warming.

By 1993 scientists had the results from Greenland. They initially thought that maybe rapid warming was local to that region, and might have something to do with changes in the Gulf Stream off Canada. So they quickly studied other regions. They examined ice sheets in Antarctica; glaciers in Peru, New Zealand and the Himalayas; the mud on the continental shelf off Venezuela, Pakistan, Baja California and US California; and ancient stalactites and stalagmites in caves in Brazil, Israel/Palestine, France and China. None of these places could provide as precise a record as the Greenland ice. But all of them showed the same periods of rapid warming as the Greenland ice cores at the same times.[6]

By the end of the 1990s the scientists knew abrupt warming was common and global. They had pieced together a reasonable picture of abrupt change in the past. In each case the switch began gradually with small changes over many thousands of years in the earth's orbit around the sun. These changes happened in regular, slow cycles over thousands and tens of thousands of years. The amount of heat and light reaching the earth stayed the same. But the angle of the light, and where it hit the earth, did change slightly. That in turn changed, slightly and slowly, the balance between the seasons and the balance of heat between the northern and southern hemispheres.

When the earth grew warmer, the level of CO_2 increased in lock step with the temperature. (So did the levels of methane and water vapour.) Somehow the rising temperature or the changing light increased the level of CO_2. That increased CO_2 further warmed the world until suddenly both the temperature and the CO_2 levels exploded. Abrupt climate change had arrived.

So the climate appears to have had two reasonably steady states—ice ages and warm periods. But CO_2 is now 100 parts per million greater than previous warm periods. This does not mean that another abrupt change is automatically going to happen. We don't know exactly when we will reach the edge of a change to another, hotter steady state.

The worry here is not that it will be unstoppable "runaway change". It is true that once we pass a certain point, the temperature will rise a good deal, and it will be very hard to slow that down. But at some point the earth will reach a new, much hotter equilibrium. Human life will still be possible at the new temperatures, although life will be much harder and quite different.

Feedbacks and tipping points

So abrupt climate change matters. But what will cause it, and when will it happen?

Once climate scientists had the data from the Greenland ice cores in 1993, they knew they were looking at some sort of feedback process. There was simply no other way of explaining what they were seeing. Somehow the warming increased the levels of CO_2, and that increased the warming, and that warming increased the levels of CO_2, and so on. Then at some point the process accelerated to the point where both warming and CO_2 levels took off.

The Scale of the Problem

A simple example of a feedback is a microphone for a rock band that is too close to the amplifier on stage. The sound goes into the microphone, out of the amp and right back into the mike, out of the amp louder, back into the mike and out of the amp louder. In a fraction of a second the noise is a scream.

That's a fast feedback. The ones that happen with climate change are measured in years, not seconds.

The climate scientists started looking for feedbacks, and for the "tipping points" where the feedbacks would suddenly accelerate.

It took some time before most scientists accepted the idea of abrupt climate change. What is surprising, though, is how quickly general acceptance came about. The first reports on Greenland were published in 1993. The people who had been in Greenland knew what they were facing by then. The majority of other scientists were coming round to the idea of abrupt change by 2000. By 2007 abrupt change was part of the consensus view in the comprehensive report of the Intergovernmental Panel on Climate Change.

Scientists have now found several possible feedback systems, and evidence that most of them have already begun. Scientists are still not agreed which ones will be crucial, or exactly when they will take off. But these are not alternative explanations—we can expect these feedbacks to work together, reinforcing each other.

Let's look now at some of the feedbacks they have discovered. I'll start with rising levels of water vapour. Water vapour—H_2O—is a greenhouse gas. It is much weaker than CO_2, but much more common. Unlike CO_2, human action has not been releasing more H_2O into the air directly. But there is an indirect effect from global warming. The warmer the air, the more water evaporates from the oceans. At the same time, warmer air is able to hold more water vapour. This increased water vapour has a greenhouse effect that further warms the air. That warming evaporates more water, which warms the air more, and so on.

Global warming has already raised the amount of water vapour in the atmosphere by a global average of 4 percent. This happens mainly in the tropics, and is one of the causes of increased intensity in hurricanes and cyclones.

Other feedbacks have also begun working. For example, the natural escape routes—"carbon sinks"—are no longer working as well as they were 20 or even ten years ago.

Until recently plants and trees on land absorbed much of the increasing CO_2 emissions. The plants could not keep up with all the new carbon in the atmosphere, but they were absorbing part of it.

The mass of vegetation on the land was growing. But there are limits to this escape route—the vegetation can only absorb so much carbon. Past a certain point, plants don't get enough water or sunlight to keep on growing more. Or the temperature grows hotter, and plants find it harder to grow. So at some point the land sink will begin to shrink.

At the same time, as the ocean warms it becomes more acidic. Plankton that absorb CO_2 find it harder to live, and the small organisms find it harder to make their shells. So the oceans absorb less carbon dioxide.

There is now clear evidence that the land and sea escape routes are absorbing less carbon dioxide. We know this because more of each year's emissions are remaining in the atmosphere. And there is a consensus among scientists that both sinks will shrink in the future as CO_2 levels in the air increase.

It seems likely that these land and sea feedbacks in themselves will only be one contributor to abrupt change. Their real importance is that, if unchecked past a certain point, they will make it very difficult for human action to stop climate change. At the moment we need to cut global emissions to the point where the carbon sinks can absorb the rest. The exact scale of how much we need to cut back I look at below. But if we do not act in time, the sinks will become less effective and the cuts will have to be greater still. Beyond that there will be a point where dying vegetation is producing more CO_2 and methane than the sinks are absorbing.

Another possible feedback arises because there is more carbon stored in the soil than there is in all the vegetation on earth. Bacteria in the soil cause carbon to decay. At moderate temperatures these bacteria multiply slowly. So more carbon is entering the soil from dead plants than leaves it. At hotter temperatures, however, the bacteria multiply quickly, the decay speeds up, and large quantities of CO_2 enter the air.

There is also a potential problem in the Amazon. A recent study by the Hadley Centre in Britain estimates that global warming may create a dieback that will turn most of the Amazon rainforest into a desert by 2100. The estimate is that this will release about 8 percent of all the carbon in soil and plants all over the world.[7] This is some time away. However, rainforests are being destroyed everywhere now. This can lead to dieback on its own, but it also releases large amounts of carbon from the soil.

The Amazon is a long term problem. The "albedo" feedback of melting snow and ice may cause abrupt change much more quickly.

The Scale of the Problem

Snow and ice are bright, dazzling white because they reflect nearly all the sunlight that hits them straight back into space. By the same token, they absorb almost no heat. But as the temperature rises in the Arctic, snow and ice melt earlier in the year and freeze later. The line of permanent snow and ice recedes towards the pole. On land, green trees and brown soil replace the snow and ice. At sea, dark ocean replaces the white ice. These new surfaces look dark because they are absorbing more sunlight and heat. So the land and the atmosphere heat up more. That in turn melts more snow and ice, and the atmosphere heats still more.

The albedo effect has already begun to work, and is one of the reasons why the Arctic has warmed far more quickly than the rest of the world over the last 30 years. James Hansen of NASA and his colleagues have recently argued, in a series of convincing papers, that the past has also seen "albedo flips"—or tipping points.[8] At these tipping points great continental ice sheets melted abruptly, both raising global temperatures and flooding low lying land all over the world.

The key places for an albedo flip now are the great ice sheets of Greenland and Antarctica, where most of the world's ice is stored. Ten years ago geologists thought that these ice sheets melted slowly. Recent careful studies have revealed that they are collapsing into the ocean far more quickly than expected. It turns out that they don't simply melt from the top down. There are also cracks that carry the warm water far down into the sheet. Aerial photographs show long fissures in the Antarctic ice near the ocean. Also it's warmer down at the base of the sheet, and the ice there turns to slick slush. The ice, honeycombed by fissures, can then slide out to sea.[9]

There are good reasons to believe that this melting may be a crucial contributor to any tipping point. One reason is that in the past climate change was much more abrupt when moving from cold to warm than the other way round. Rapidly cracking ice sheets seem to explain this discrepancy—they only happen when the earth is warming, not cooling. Moreover, the computer simulations by Hansen and his colleagues show that this scenario fits earlier occurrences of abrupt climate change. And it is not necessary for the ice and snow to melt right down to the ground or the open ocean. It is enough if the surface snow and ice melt and are replaced by light blue water.

Another serious feedback has begun with methane in the soil in the far north. As I mentioned earlier, methane is a particular worry for abrupt change because it has such a powerful effect during its first few years in the atmosphere. This means that large amounts of methane can trigger swift and enormous feedbacks.

At present very large areas of Siberia and Canada are covered by frozen tundra. The far north, as we have seen, is melting much more quickly than the rest of the world. Much of this tundra is peat, which contains considerable reserves of methane. When peat thaws, it becomes soft and spongy, as anyone who has ever tried to walk across a peat bog knows. This means that the peat easily releases the stored methane. The methane then warms the air further. This melts more permafrost, which releases more methane, which melts more permafrost, and so on. This process has already begun over large areas of Siberia.[10]

All the feedbacks mentioned so far are ones that have already begun. One further feedback mechanism—the melting of "methane hydrates"—could be very serious, but has not yet begun. Very much larger amounts of methane are stored on the ocean floor in frozen crystals of methane and water. The crystals clump together into large deposits. These "methane hydrates" are found around most continents at depths of 500 metres or more. The cold of the deep ocean and the weight of the water above keep these hydrates frozen in place. But when a deposit of methane hydrates warms, it melts and bubbles to the surface in "methane burps". There is evidence that this has happened often in the past. The deep ocean pressure on the hydrates is such that a deposit of one million litres of hydrates under the sea can suddenly become 160 million litres of methane when it hits the air. Moreover, the gas hydrates of the ocean contain about twice as much carbon as the total amount in all the remaining coal, oil and gas in the world.[11]

Methane hydrate deposits are potential global warming bombs. If methane hydrate burps to the surface it could warm the nearby air almost instantly, thus melting more hydrates, thus warming the air more, and so on. The bad news is that gas hydrates are most common in polar regions, and the Arctic is already warming much more swiftly than the rest of the world. The good news is that the hydrates are deeply buried. It will probably be some time before many of them melt, and it may never happen on a significant scale. In short, methane melting out of the permafrost will have a serious effect soon. The melting of methane hydrates from the ocean would be catastrophic, but scientists are unsure whether they will happen on any significant scale.

There are more feedbacks, and more will be discovered. These feedback processes will combine, in both predictable and unforeseen ways.

The Scale of the Problem

How much do we need to cut?

Abrupt climate change, then, is a serious threat. We know it happened in the past, and that feedbacks and tipping points were involved. We know that if we keep forcing the pace of climate change through our own carbon emissions something of the sort will happen again. We know that several different feedbacks have already begun to happen.

It is clear abrupt climate change will happen if we don't act. What is much less clear is the timing at which we can expect abrupt climate change of any sort to take off. There are two problems here. One is that scientists do not yet know which feedback will be the key driver of abrupt change. The other is that predictions about future climate change are made using computer simulations.

The global climate system is very complex. Computer simulations work by developing a model of this system. The different models are really sets of equations predicting relationships between, for instance, carbon dioxide levels and temperature, or temperature and water vapour, or the speed at which glaciers of a certain size are likely to melt. The predictions from the simulation are then tested by "hindcasting". The computer simulation is run to see if it fits with what actually happened in the past. If it does not, the equations are reworked until the fit between the model and actual events in the past is better.

One difficulty here is that the global climate system is far more complex than any simulation. This means that different computer simulations at different universities produce a range of predictions. But the real difficulty with predicting abrupt change has to do with the equations in the simulation. These equations assume that the relationships between the different factors are constant and stay the same. Abrupt climate change, however, happens when the fundamental relationships between temperature, water vapour, CO_2 and so on change. The computer equations, by their very nature, are bad at predicting such fundamental changes.

So predictions of the point at which abrupt climate change is likely to start are only educated guesses. However, there is now quite general agreement among scientists that serious abrupt change will kick in with an average global temperature rise of somewhere between 2.0°C and 5.0°C since 1800. Towards the top end of that range abrupt change is very likely indeed. There is also general agreement, but not complete, consensus that abrupt change might well begin at 2.0°C and it would be wise to hold temperature rises to that limit.

Temperatures have already risen 0.7°C since 1800. So 2.0°C in fact means a further rise of 1.3°C on present levels.

It would be much better if we could be more precise than that. The fact is we can't, and whatever temperature rise we pick is to some extent arbitrary. Moreover, the arbitrariness is not just scientific. Whatever figure experts pick for the onset of abrupt climate change is also heavily influenced by political pressures.

For example, in 2006 Sir Nicholas Stern wrote a long authoritative report on climate change for the British government.[12] Stern is a mainstream economist and was then a senior civil servant in the British Treasury. He wanted to persuade his political masters that they had to take decisive action on climate change. But he also wanted to recommend action that his masters might actually take. So he warned of increasing risks of abrupt change as the average temperature rise went from 2.0°C to 5.0°C. He then had to choose the lowest level at which he thought it would be safe to aim for stabilising CO_2 levels. He chose 495 parts per million of CO_2. This level will take us well over a temperature rise of 2.0°C.

The Hadley Centre, the research wing of the Meteorological Office in Britain, is under less direct pressure, and has estimated that a rise to 2.0°C is as far as we can go. The European Union officially went for this level as well in 2007, and there seems to be a reasonable consensus among European scientists that this is the safe limit.[13] However, it is important to understand that this does not mean abrupt climate change will automatically begin when we hit a rise of 2.0°C. It means abrupt climate change starts to become likely at 2.0°C.

What all this means is that the best *estimate* is that a level of between 400 and 450 parts per million of CO_2 will produce a temperature rise of 2.0°C. So we will reach the danger point for abrupt change with another 15 to 65 parts per million of CO_2 in the atmosphere.[14]

With currents rates of CO_2 emissions and the continued operation of the carbon sinks we will probably, but not certainly, avoid abrupt climate change if we stabilise the amount of CO_2 in the atmosphere in the next seven years. We may be all right if we do it in 31 years. To simplify matters, for the rest of this book I will say that we have between ten and 30 years in which to act. It is also possible, but not that likely, that we have longer. And it is possible, but unlikely, that we have already passed the tipping point.

To stabilise the amount of CO_2 in the atmosphere, we have to cut the amount we put into the air each year. We have to reduce it to a

The Scale of the Problem

level where the carbon sinks, the natural escape routes, can take up all the CO_2 we are putting into the air. Then the amount of CO_2 in the air will stay steady. That means we need to reduce annual emissions to 1.4 parts per million. That's a cut of 60 percent from the current rate of annual emissions.

However, that will only work if we act quickly. As we have seen, when the total level of CO_2 in the atmosphere rises, the carbon sinks will work less well. That may mean we have to cut emissions by up to 70 percent.

In other words, to have a reasonable chance of avoiding the horrors of abrupt climate change, we need to cut global emissions by between 60 percent and 70 percent. And we have somewhere between ten and 30 years to do that. That is the scale of the political task ahead of us.

Chapter 2

Poor People are not the Problem

THE last chapter argued that the threat of abrupt climate change means we need cuts in carbon dioxide emissions of 60 to 70 percent. Those are deep cuts, but they are possible. This chapter will argue, however, that to allow for economic growth in the poor countries, and for population increases, cuts in the richer countries will have to be in the order of 80 to 90 percent. This is still possible, but will require radical change in the global economy.

This chapter also takes on the argument that blames the global poor for climate change. This argument, now common, says there is no point in even trying to cut carbon emissions because India and China are growing so fast that what "we" do in the richer countries won't make any difference.

This argument for despair serves the interests of those corporations and governments who want to do nothing. This chapter shows that the argument is simply wrong about the numbers. I will show that, even with economic and population growth in the poor countries, it will still be possible to cut global emissions enough to avoid abrupt climate change.

The standard argument

Many people feel torn, because they want to do something about climate change and they want to make poverty history. But their understanding is that they can't do both. They have to choose. That makes them feel helpless. But we need to be clear about who is pushing this argument on a global level, and why.

There is now a global debate about what kind of international climate treaty will follow Kyoto in 2012. There are basically three positions in the debate. One is the position of environmental campaigners and climate activists. We want a much stronger and more effective international treaty. The second is the position of most governments and politicians. They want a new international treaty with deeper cuts in carbon emissions, but one that does not go too far or ask too much of corporations.

The third position wants no international agreement at all. At the moment the leadership for this position comes from George Bush's US administration. They speak for the oil companies, the coal companies and the car companies. But their anti-treaty arguments also rely on the support of the governments of India and China. As I will argue in the next chapter, the Indian and Chinese governments do not represent the interests of their people any more than Bush represents the interests of ordinary Americans. All three governments want to avoid action on climate change for similar reasons.

In 2005 George Bush could simply deny that climate change was happening. That is no longer possible. Now those who want to do nothing say that there is no point in a stronger international agreement after Kyoto, because India and China will not go along with it. Instead Bush and his like say "we" have to involve China and India in a different kind of international agreement. Unlike Kyoto, such a new agreement will not ask for cuts in carbon dioxide emissions. Instead it will fund massive investment in research to discover new, painless ways to limit CO_2 emissions. While the world is waiting for the research to be done, emissions can keep on growing.

The government of the United States says they would like to do more, but they can't, because India and China won't. The governments of India and China say they would like to do more, but they can't, because they have all these poor people.

Everyone blames the poor for policies that will devastate the lives of the poor. Their standard argument goes like this:

The Chinese economy is currently growing at more than 10 percent a year, and India almost as fast. China has 1.3 billion people and India one billion.

China is perhaps already, or will soon become, the largest carbon dioxide emitter in the world. Worse, within 20 years China will be as rich as the US is now. Then the Chinese will emit as much CO_2 per person as the Americans do now. That will be 28 billion tons a year—as much again as the whole world emits now.

India is poorer than China, but when it too reaches American levels it will be emitting 23 billion tons.

Moreover, the population of the world has increased from one billion in 1800 to 6.5 billion now. If the people in poor countries keep having babies, there is no way we can stop global warming.

*What makes it worse, India and China will not give up on eco-
nomic growth. Holding back emissions in the rest of the world
won't make any difference if they don't cooperate. And they won't.*

The basic flaw in the "blame the poor, blame India, blame China"
argument is that its proponents assume that India and China will
develop to be exactly like the United States. This ignores the fact
that India and China could as easily develop to be more like Europe
than the US, but still as rich.

I will take the numbers on economic growth first, and then the
numbers on population.

Up to now I have been talking about global CO_2 emissions as a
proportion of the total atmosphere—3.5 parts per million more a
year. From here on we will be talking about the emissions for each
country and each industry. Parts per million are a clunky way of
measuring those emissions. So from here on I will measure emissions
in tons of carbon dioxide.

One part per million is roughly equal to eight billion tons of CO_2.
Total global emissions of CO_2 each year are now about 28 billion
tons. The sinks take up about 11 billion tons, and 17 billion tons
stay in the atmosphere each year. The last year for which we have
reliable statistics is 2004. That year global emissions were slightly
less—27 billion tons. The average global emissions for each of the
6.5 billion people on earth were 4.2 tons per person.

The different rates of CO_2 emissions per person for different
countries in 2004 are very telling. The first row in the table below is
emissions of CO_2 per person in tons. The second row is total annual
emissions in billion tons of CO_2.

*Annual total CO_2 emissions and emissions per person by country,
2004*

	Tons per person	Total billion tons
GLOBAL	4.2	27.0
USA	20.2	5.9
Canada	18.1	0.6
Russia	11.7	1.7
Japan	9.9	1.3
Germany	10.5	0.9
UK	9.6	0.6
Spain	9.0	0.4
France	6.7	0.4

The Scale of the Problem

EUROPE	8.0	4.7
Italy	8.4	0.5
Iran	6.4	0.4
China	3.6	4.7
Central & South America	2.4	1.1
Brazil	1.8	0.3
AFRICA	1.1	1.0
India	1.0	1.1[15]

The European average is 8.0 tons per person. The US average is 20.2 tons per person. That's two and a half times the European average, and double the average for Germany, Japan and the UK.

If China develops to become as rich as the United States, with an economy just like the United States, it will produce more than five times the emissions it does now. If China develops to become as rich as Europe, with an economy like Europe's, it will only have a bit more than twice the emissions.

But would it be possible for China to develop on an European pattern? It is often said that China is very inefficient in energy use and that Chinese industry is wasteful on an American scale. This is true if you look at the figures for CO_2 emissions and gross national product in dollars. Dollar for dollar, China emits a great deal of CO_2. But these figures are misleading. They are based on a fixed exchange rate that makes Chinese goods much cheaper than American ones. This exaggerates Chinese poverty, for Chinese incomes are measured in terms of what they can buy in America. In fact they can buy a great deal more in China.

If you divide the annual CO_2 emissions from China not by dollars, but by the things China produces, then China is still slightly above the global average.[16] Much of this difference is because China relies heavily on coal for electricity. But China has vast reserves of sunlight, and particularly of wind in the west. So it should be perfectly possible for China to develop along similar lines to Europe.

We can't stop global warming if China and India develop on the American model. We can if they develop on the European model. It will take a bit of arithmetic to explain why. I will use the Italian model as my example. Average Italian emissions (8.4 tons per person) are exactly double the current global average (4.2 tons per person).

Let's imagine—this isn't going to happen, but let's imagine it— that in the next 20 years the whole world becomes as rich as Italy and emits as much CO_2 as Europe. That would mean not just China and India, but Bangladesh, Pakistan, Vietnam, Africa, Latin America and all the rest. And let's imagine that the heavily polluting countries like the US get their emissions down to Italian levels.

In that case, all the countries in the world would be emitting an average of 8.4 tons of CO_2 per person. Right now we are emitting a global average of 4.2 tons per person. So with economic growth for all the poor countries, total emissions would double.

But remember, we need to cut the total of global emissions from their *present level* by between 60 percent and 70 percent. Right now that means a cut from 4.2 tons of CO_2 per person to between 1.7 and 1.3 tons per person. If the whole world was on an Italian level of 8.4 tons of CO_2 per person, we would still need to cut emissions down to between 1.7 and 1.3 tons per person. That would mean cuts of between 80 and 85 percent.

In short, we can achieve this imagined world where poverty is history and still stop climate change. It would need cuts of 80 to 85 percent per person in the rich countries. This sounds impossible. But it can be done. I will show how in the next chapter.

However, there are two unrealistic assumptions in this imagined world I've just described. The first is that all the poor countries on earth will become as rich as Italy by 2027. There are strong reasons to believe Indian and Chinese growth rates will fall. No country has ever had rates above 5 percent over the long haul, and growth rates usually drop once countries became industrialised. But even if India and China continue to develop at their present pace, no one thinks that Bangladesh or Sudan will. So cuts per person may have to be less than 80 percent in the richer countries.

The second unrealistic assumption is that the United States, Canada and Australia will make much deeper cuts than anyone else. In my imagined scenario, the US has to reduce to Italian levels, and then cut 80 percent or more from there. This is a total cut of 92 to 94 percent. This is possible in the very long term, but unlikely by 2038. Slower American cuts in emissions, however, will be balanced by slower economic growth in the poor countries.

In any case, the trade-offs between economic growth and emissions cuts in China, India, the US and the other countries will actually be a matter for complex and difficult conflicts and negotiations on a global scale. My purpose here is only to show that it is possible to reduce emissions to a safe level and still make poverty history. Poor people are not the problem.

The Scale of the Problem

Population growth

That's the argument for economic growth on the numbers. The people who want to blame China and India for global warming also use the wrong figures for population growth and the climate consequences of population growth.

In fact, as I shall show below, likely global population growth will only increase emissions by 10 to 20 percent over the next 25 years. My argument here will almost certainly be unfamiliar, but it is sound. We can halt global warming with cuts of 60 to 70 percent now. We can also make room for both population growth and a richer world for everyone with bigger cuts of 82 to 87 percent over the next 25 years.

The "blame the poor, blame China, blame India" argument I quoted above has a corollary about population and climate change. It usually goes like this: *Global population has tripled in the last century. If it triples again in this century, all those new people will flood the skies with carbon dioxide.*

Luckily, the facts are at odds with these fears. Global population is on course to go down.

It's true that global population has increased six-fold since 1800. This rise has not come from people having more children. It has come from fewer children dying and adults living longer. But this does not mean population will continue to grow in the same way. Instead something unexpected is happening. In most countries the rate of population growth is slowing down. These countries include the vast majority of humanity. In all the rich countries, and some of the poor ones, women in the present generation are having so few children that the population will soon fall. The trend is down, and the earth's population will begin shrinking at some point this century.

To see this, we need only some simple arithmetic. If each woman has an average of two children, a boy and a girl, over the long term the population of the world will stay the same. In practice, though, some baby girls do not survive to adulthood and not all women have children. With current death rates, the global population will stabilise if the average adult woman has 2.1 children.

Of course, the global population will not fall immediately. Older people are still living longer than they used to. The population will only stabilise once the average age of death levels out. This may take 30 to 50 years, as the people now in their teens and twenties reach old age. However, in the countries with the lowest birth rates the population totals have already begun to fall.

Here are the numbers for average children per woman in different countries now.[17] Remember, below 2.1 children per woman the long term trend is down. The numbers are unexpected:

Children per woman

Hong Kong	0.9
Taiwan, South Korea	1.1
Singapore, Bosnia	1.2
Italy, Spain, Germany, Russia, Bulgaria, Romania, Hungary, Poland	1.3
Switzerland, Portugal, Croatia	1.4
Cuba, Canada, Cyprus, Netherlands	1.5
Trinidad and Tobago, Georgia, Belgium	1.6
(1.6 is the average for all developed countries)	
Thailand, Armenia	1.7
Australia, United Kingdom, Denmark, Puerto Rico, Serbia, Mauritius	1.8
Albania, Costa Rica, Ireland, France	1.9
USA, Sri Lanka, Iran, Tunisia, North Korea, Azerbaijan, Chile, New Zealand, China[18]	2.0
Vietnam, Iceland	2.1
(2.1 is the break even point for population growth)	
Turkey, Uruguay, United Arab Emirates	2.2
Brazil	2.3
Indonesia, Mexico, Argentina, Colombia, Kuwait, Lebanon, Algeria	2.4
Malaysia	2.6
Venezuela, Panama	2.7
(2.7 is the average for the world)	
South Africa	2.8
India	2.9
Bangladesh	3.0
Egypt	3.1
Philippines	3.4
Pakistan	4.6
Haiti	4.7
(5.5 is the average for sub-Saharan Africa)	

Several things here run counter to the usual assumptions of the mainstream media. Religion seems to make no difference to birth rates. Many Catholic countries, like Italy, Spain, Poland and Portugal,

The Scale of the Problem

have among the lowest rates of population growth in the world.[19] Southern Europe has lower rates than Northern Europe. The Islamic religion seems to make no difference. Indonesia, the world's largest Muslim country, is on 2.4. Iran is on 2.0, the same as the USA, a much richer country.

Richer countries tend to have lower rates than poorer ones. Almost all the very poorest countries have high birth rates. But several "less developed" countries have below 2.1 children per woman, including Cuba, China, Trinidad, Thailand, Armenia, Mauritius, Costa Rica, Sri Lanka, Iran, Tunisia, North Korea, Chile and Azerbaijan.

There is an enormous literature on the reasons for the fall in population growth, but the key is economic security. In quite different cultures around the world, people traditionally had a number of children because there was no other social security provision for accidents, illness or old age. With high mortality rates, they could not know how many of their children would survive. Once people have decent medical care for children and a welfare net to protect them from extreme poverty, they begin to limit their children.[20]

Population growth in the future

These population figures have important implications for carbon dioxide emissions and climate change. To repeat, the world's population is increasing slowly. It will level out and then fall. But the really important question here, as with all carbon dioxide emissions, is not what will happen in the poor countries of the world, but what will happen in the rich countries. These are the countries that use more energy and so produce more emissions.

This is a very important point to get straight. People sometimes talk as if a 50 percent increase in global population will mean a 50 percent increase in emissions. It won't. Doubling the population of Bangladesh by another 140 million people would increase global emissions by .001 percent. Adding 140 million people to the 300 million in the USA would increase global emissions by over 10 percent. That's more than 1,000 times as much emissions as from the same population increase in Bangladesh. The danger isn't poor brown people having babies. It's rich white people having babies. And luckily they aren't.

The average woman in a developed country now has 1.6 children. This will soon bring the population down, even if there is substantial immigration from poorer countries.

Here are the United Nations estimates for likely population growth in the richer countries between 2005 and 2030. As I have said, global population will begin to fall at some time this century. But my focus here is on the period between now and 2030, because we need to act on climate change by then.

The UN figures use three estimates for population growth—a high estimate, a medium one and a low one. I will use the low and medium estimates, for reasons explained in the footnote.[21]

Population growth, all more developed countries

Total population, 2005   1.22 billion (estimate)
Total projected population, 2030
   Low estimate   1.18 billion (3 percent fall)
   Middle estimate   1.26 billion (3 percent increase)[22]

In other words, the richer countries will see either a slight fall or a slight rise in their populations. The situation with global population is different, but still manageable:

Population growth, all countries

Total population, 2005   6.5 billion (estimate)
Total projected population, 2030
   Low estimate   7.7 billion (18 percent increase)
   Middle estimate   8.3 billion (28 percent increase)

In other words, population in the rich countries will stay roughly steady. All the increase will be in the poor countries. The global increase will be between 18 percent and 28 percent. Let's look at the effect of that population increase if nothing is done to cut emissions. The effect of increased population on CO_2 emissions in the rich countries will be about 0 percent. In the poor countries, it will be closer to the 18 percent to 28 percent increase in global population. Taking the two together, a reasonable guess for CO_2 emissions due to population growth would be 10 to 15 percent.

The implications for emission cuts

Now let's look at what an increase of emissions of 15 percent due to population growth would mean for the cuts we have to make. I

                                             The Scale of the Problem

have argued that to make the whole world as rich as Italy and stop climate change we would have to cut emissions per person in the rich countries by 80 to 85 percent.

If we allow for, in effect, 15 percent more people, the total cuts would have to be 83 to 87 percent. (The arithmetic is in the footnote.)[23] This would allow for ending global poverty and for any expected population increases.

However, those figures—83 to 87 percent—are probably higher than they need to be, because they assume that all the poor countries will become as rich as Italy in the next 25 years. With any of the rates of economic growth we are actually likely to see, emission cuts of 80 percent per person in the pattern of emissions in the rich countries will be more than enough to cater for global development and population growth.

To sum up, we can allow for any expected economic growth and the expected population growth, and still only have to cut emissions by roughly 80 to 85 percent in the richer countries.

CHAPTER 3

Sacrifice is not the Answer

THE last chapter looked at the numbers and showed that poor people are not the problem. Now I turn to the politics of poverty, and stand the usual argument on its head.

The usual approach to the "problem" of India, China and climate change starts from the question of how "we" can cope with economic growth in the poor countries. It assumes that "India" and "China" are blocks of people, and that the interests of Indian and Chinese people are the same as their governments and corporations. And it assumes that the only solution to global warming will be an end to economic "growth", and require "sacrifice" by ordinary people in Asia *and* Europe.

I am going to show you how the world looks if we discard this set of assumptions. Here I will return to the point I made in the introduction. Serious action to stop abrupt climate change will not mean sacrifice by ordinary people. It means government intervention, regulation, massive global aid, and hundreds of millions of jobs.

"We"

Most writers on global warming make a silent assumption that "we" are the people in the rich countries. They write, for instance, that "we" are prisoners of consumerism, buying far more than we need. Some books assume that "we" are only Americans. They write that burning less oil will be good for "our" security because it will cut "our" dependence on Middle Eastern oil.

Moreover, in American and European books, "we" almost always really means educated, affluent professionals whose income is higher than 80 or 90 percent of the population. As used in American social science books, for instance, "we" usually does not include immigrant Hispanic waitresses in California, white Anglo carpenters in Vermont, former steel workers serving in Pittsburgh Pizza Huts, slaughterhouse workers in Iowa, truck drivers in Chicago or prisoners in Texas. Built into "we" is an identification with the affluent in one's own country and a disdain for both foreigners and the majority of one's fellow citizens, who become "them".

This book starts from a different point of view. I write as a citizen of the planet, for other citizens of the planet. My "we" embraces myself and other ordinary people who work with our hands, hearts and brains. That "we" includes the great majority of humanity.

So I shall not concern myself with what "we" do about "them" in India and China. Half the people in the world live in Asia. I start from what ordinary people in Asia need. Moreover, physics makes global warming a global problem. CO_2 can be put into the air anywhere. Within two years it is evenly mixed into the atmosphere all over the world. This means that local solutions alone will not, and cannot, work.

Climate change in India and China

How does climate change look from the point of view of ordinary people in China, India and the other poor countries? Both the people and the governments want to industrialise. No one wants to be stuck in poverty. Historically, in fact, every country that has grown rich has done so by building factories and industry. People in poor countries also want to travel. Most of them want to live in cities. They do not want to be so cold or so hot that they put their lives in danger. They don't want to live six people to a room. They don't want to see their children hungry or watch them die without medical attention. They don't want to drink dirty water.

In addition, it is grossly unfair for the governments of the rich countries to tell poor countries they must not industrialise because "we" have to save the climate. It is not just that the energy use of people in poor countries is creating less than half the problem right now. There is also the climate history of the last 200 years to be considered. In those years the developed world built wealthy economies by pouring carbon dioxide into the air. China now accounts for just over a fifth of global emissions and India for less than a twenty-fifth. But if you add up emissions over the last 200 years, only a tiny part of global warming comes from India and China. The rulers and ordinary people do not think it is fair to shut the door of industry in their faces after other countries have become rich.

In India, in particular, most activists say that global warming is a Western problem. It has been caused by rich countries. Let them solve it, the argument runs, and let them pay for it. Our job is to get India out of poverty.

This is an attractive argument in India. But it's a political mistake, because it does not actually start from the needs of poor people in Asia. Global warming will devastate them. The majority of Indians still make a living from agriculture. The urban economy depends on farmers making enough to feed the cities and buy goods from the cities.

Climate change is likely to have four major effects in India. First, the pattern of the monsoons will change. Most farmers depend on these monsoon rains, and without them crops will decline and in some regions fail. Second, many farmers depend on the rivers that come from melting snows and glaciers in the Himalayas and Karakorum mountains. Those glaciers are melting. When they are gone, the rivers will begin to dry up. Third, hundreds of millions of people in India live in low lying coastal areas that will be flooded by a combination of rising sea levels and intense storms.

Finally, much of India is hot, and city dwellers die in heat waves every year before the monsoon. The most vulnerable are those who must work in the heat, like rickshaw drivers. With global warming the death toll will rocket.

Hundreds of millions will be in serious want. Many of them are very poor to begin with. There is already a thin line between them and desperation, and their government will spend little to help them. When climate change sends the richer farmers to the wall, the landless agricultural labourers will lose everything.

China is broadly similar, although somewhat richer and more unequal. There are fewer agricultural labourers, but almost half of the population are still farmers. Some in the Chinese cities are growing rich, but much of rural China remains very poor. The land is not as dependent on the monsoon, but the glaciers of the Himalayas matter to China, and many Chinese people live on the coast.

Global warming threatens not only ordinary people in India and China, but also many of the poorest on earth. The heavily populated places most vulnerable to flooding are the river deltas of the Nile in Egypt, of the Ganges and the Brahmaputra in Bengal and Bangladesh, and of the Mekong in Vietnam. Drought has already helped to create famine and war in the African Sahel. In short, poor people in poor countries have more reason to stop global warming than anyone else on earth. But they also need economic growth to escape from poverty.

The Scale of the Problem

"India" and "China"

What kind of economic growth do ordinary people need?

Most writing on poor countries and climate change starts from an implicit assumption that there is one united actor on the world stage called "India" and another called "China". The interests of Indian and Chinese people are assumed to be the same as their governments'. So people say and write that "India wants this" or "China insists on that".

Few people make this assumption about the United States or Britain. Climate activists in the US are painfully clear that there is a difference between what they want and what George Bush and the oil companies want.

To make sense of the inequalities in China and India, we have to understand that they have a ruling class, just as Western countries do.

In Western countries government policy is driven by the interests of the ruling class. That ruling class includes the leading executives in governments, corporations, banks, media, the armed forces, the foundations and the universities. These powerful people run the world in the interests of the rich. This ruling class is the people who make the decisions. There are different ways of defining the boundaries of this class, but however you do it not many people belong to this elite.

Below this ruling class are a much larger number of middle managers, professors, police chiefs, colonels, stockbrokers, consultant doctors, editors, bishops, small business owners and the like. These people do not run the world, but they do make good salaries. They administer the world in the way the ruling class tells them, at least most of the time, and provide the ideas that justify the status quo.

The great majority of people in every country have neither wealth nor power. These are the people who scrape a living on farms or work for wages and do what they're told every working day. They include both manual workers and white collar workers who also make ordinary salaries and do what they're told. Most of the time white collar and blue collar workers accept much of the world view of the ruling class. After all, those are the dominant ideas in every society. But working class people have another set of ideas in our heads at the same time. These emphasise sharing, kindness and equality. Without these values we could not build friendships or raise children or get through a day at work. By and large, most of us feel we have to accept the world as it is. But we also have a private view of how the world should be. When ordinary people build their own unions and political parties, those organisations reflect those decent values.

The ruling class rule the world, but they don't have it all their own way. There is a constant struggle about the shape of every society. This is clear enough in democracies. It is also true in dictatorships, although under more difficult circumstances. Everywhere there are debates and fights over pensions, healthcare, working conditions, minimum wages, housing, land rights and environmental regulations. At the core of these fights are two truths. One is that ordinary people vastly outnumber the rulers. The other is that the ruling class has far more power. Shifting compromises are the result, in most times and places, between what the ruling class wants and what ordinary people need. The balance of these compromises favours the ruling class, but they do always have to compromise.

These class divisions exist in North America and Europe. They also exist in Latin America, Africa and Asia. There too the rich send their children to Harvard and Oxford. The divisions between rich and poor are of the same order as everywhere in the world. But the divisions bear more acutely in the poorest countries, where poverty has terrible consequences.

It is important to be clear on the reality of ruling classes in poor countries because of their effect on the politics of climate change. The current American government does not want any global agreement to reduce CO_2 emissions. To make sure it does not happen, they are counting on the support of the governments of China and India. Those governments currently support George Bush's plans. They do not do so because it is in the interests of ordinary Asians, whose lives will be devastated. They do it for the same reasons George Bush and American corporations do it. They too want to compete in the global economy, to be rich and powerful.

Unequal economic growth

Now let's look at the politics and economics of climate change in India and China in terms of what the rich there want and what ordinary working people need.

Commentators usually write as if economic growth in poor countries benefits everyone equally, as if when a country builds industry and becomes rich, the people do too. But the reality is different. In both India and China economic growth in the last 20 years has gone along with growing inequality. The people who benefit least from economic growth are the people who will suffer most from climate change.

Makesh Ambani, the billionaire oilman and head of Reliance Industries, is building a skyscraper in the centre of Mumbai as a home for his extended family and no one else, except their 600 servants. The building will have 27 floors, but with ceilings so high it will be as tall as a 60 storey office block. The point is in your face flaunting of wealth.

Much of Indian economic growth is driven by production of luxury goods for the "new middle class". These are the people in the new cars, taking the new cheap airline seats, crowing about the success of "India shining". Yet about one third of rural people are still agricultural labourers, and their already desperate circumstances are becoming worse as they lose jobs to tractors and pumps.[24]

In China the growing wealth and the soaring urban skylines are real too. But the economic miracle is driven by two things. One is exports. The largest markets are other countries in Asia, Europe, and the United States, in that order. The other driving force, as in India, is production for the rich minority—the Chinese "new middle class" who drive the cars and live in apartments in the new skyscrapers.

Some Chinese farmers, particularly in the rich agricultural lands near the coast, have also benefited greatly. The majority have not. Farmers used to depend on the government communes for minimal free education and free healthcare. The economic reforms have taken these away.

In Chinese cities the state used to own all the factories. State sector workers had the "iron rice bowl". This was a job for life, with guaranteed food, education, healthcare and pensions. In addition, workers in many factories had organised to win shorter hours and a gentler pace of work during the Cultural Revolution in the 1960s. In the "reforms" of the last 30 years many of these factories have been closed down. The new factories are owned by provincial and local governments, private firms and foreign companies.

It is true that there is a real "trickledown" effect from the booming economy of the rich in China. Ordinary people do have more factory jobs. In China many of them are girls and boys from poor rural families. Under Mao the government would have arrested them for leaving their villages. But these new factory workers have no iron rice bowl. Their wages do not pay enough to support a family. They sweat 14 hours a day for low wages in unsafe factories, and live packed in dormitories on the factory grounds.[25]

Of course India and China were unequal societies before the reforms. But with the kind of economic growth they are seeing now, they are becoming more unequal. The same thing has been happening in the last 30 years all over the world.

Ordinary people in India and China are not stupid. They see what is happening, and resent it deeply. India is full of activists, strikes and demonstrations. In China society is deeply split. Out of 1.3 billion people, only 60 million have a Western standard of living. Those wealthy 5 percent are delirious with happiness about the new China. But the political police also count every public protest in China. They counted 58,000 protests in 2003, 74,000 protests involving three and a half million people in 2004, and 87,000 in 2005. Most of them are organised by farmers defending their land against confiscation by business and government, people protesting against government corruption and workers defending their pensions and working conditions.[26]

China is also one of the most polluted countries in the world, because the government and the corporations use the earth with the same ruthlessness as they use their working people. But ordinary Chinese know this and hate it. In 2007 the Pew Centre did one of their periodic global surveys of public opinion. Among other things, they asked people to name what they saw as the top threats to the world. 70 percent of Chinese people—the same percentage as Japanese and only exceeded by South Koreans—chose "environmental problems" as a top threat. Ukrainians came next, with 57 percent. (Chernobyl is in Ukraine.)[27]

In 2007 the Chinese government insisted on extensive cuts from a World Bank report on pollution. The two things they particularly wanted suppressed were the estimate that 650,000 to 750,000 Chinese die every year from air pollution, and the map that showed most of the dead were in the coal mining regions. The Chinese government did this, an "adviser" told the *Financial Times*, because they were afraid of "social unrest".[28]

Of course ordinary people in China, and in India, want to escape from poverty and fear. But the growth they need is not the growth they are getting. And they do not need global warming.

The Scale of the Problem

I have just stood the ideas "we" and "India" on their heads. I now do the same with "sacrifice".

Let me begin with the question why ordinary people in the richer countries should put up with deeper cuts so that people in India and China can escape poverty.

There are some simple answers. People in richer countries are going to suffer from climate change too, and ordinary people are going to suffer more. To stop global warming, ordinary people in poor countries will have to ally with ordinary people in rich countries. It works the other way round too. People in rich countries need an alliance with people in poor countries. It just won't happen otherwise.

Moreover, ordinary people in the richer countries have a moral sense. They don't think it's right to leave people in desperate poverty.

But there's another answer to the question too. It has to do with sacrifice. Let me start with a personal story. In 2003 I was at the World Social Forum in Porto Alegre, Brazil, with 100,000 other global activists. One of our slogans was "Another world is possible". I led a workshop about what that other world might look like. Twenty people came to the workshop, mostly young men from Brazil and Argentina. We debated the state of everything, serious but having fun. Then a young Canadian spoke. She said:

> What's wrong with our society now is that people are prisoners of consumerism, oppressed by desire for things. We are using up the resources of the earth far faster than we can replace them, with no respect. We live in an economic system that glorifies greed. We have to stop going after growth, and value the quality of life instead.

In Canada what she said would have sounded like radical politics. But as she spoke I could feel the Brazilians and Argentinians growing hostile. When she finished speaking, they launched into her. Her face crumpled.

They had heard another rich *gringa* telling them to sacrifice. The Argentinians were angriest. A year before our workshop an IMF sacrifice package had produced financial crisis in Argentina, threatening jobs and savings. Two million people packed the streets of Buenos Aires, banging pots and pans all night. Those demonstrators overthrew the government. The Argentinian people went on to overthrow the government that replaced it and the next one too. Back in

1975 Argentina had been a rich country, on a par with Italy. By 2003 it was a poor country. The young men at the workshop and their parents had experienced and fought sacrifice. They heard the Canadian siding with the IMF.

Two years later, in January 2005, the World Social Forum returned to Brazil. I went to all the meetings about climate change to gather support for global demonstrations. I would begin speaking, and the mostly Latin American audiences would listen with interest. Then I would say that climate justice must also mean a global movement to lift all the world out of poverty, and suddenly they applauded.

I learned a simple truth there. People in poorer countries will not back action against climate change if it is presented as sacrifice. But ordinary people in richer countries don't want sacrifice either. In the last 30 years most working class people in Britain, for instance, have had attended a meeting at work or in the neighbourhood, where a man or woman in a suit comes from the employer or the council. The suit explains to them what sacrifices they will have to make for the good of the company or the country. Working people have learned that speech foretells the future. They will sacrifice, and the suit will not. This has been bitter experience, taken to the bone— and not just in Britain.

Moreover, what working people everywhere worry about most is being asked to sacrifice their basic security. An ecological campaigner may talk about giving up expensive clothes, but a working class audience hear they will lose their jobs. The campaigner says we have too much. Ordinary people feel—*we don't have enough*.

If you are on an average income anywhere in the world, look at what you spend during the month. At least 80 percent of most people's budgets goes on food, clothes, housing, utility bills, transport to work, the children and one night out a week. For many, these basic expenses take up over 100 percent of their income and they live on debt.

Of course ordinary spending does include some things people don't need. But there's a reason people want those things badly, and it isn't greed. "Things" show where you stand in society. The socialist writer on health Vincent Navarro puts it this way:

> An unskilled, unemployed, young black person living in the ghetto area of Baltimore has more resources (he or she is likely to have a car, mobile phone, and TV, and more square feet per household and more kitchen equipment) than a middle class professional in Ghana,

The Scale of the Problem

Africa. If the whole world were just a single society, the Baltimore youth would be middle class and the Ghana professional would be poor. And yet, the first has a much shorter life expectancy (45 years) than the second (62 years). How can that be, when the first has more resources than the second? The answer is clear. It is far more difficult to be poor in the United States (the sense of distance, frustration, powerlessness, and failure is much greater) than to be middle class in Ghana.[29]

That person in Baltimore may not need those extra things, but he or she does not want to give them up. They are a sign that at least they have something. Possessions are the way people in their society keep track of power and powerlessness.

Building a movement

In the next chapter I will begin to lay out, in detail, the sort of measures that can stop global warming without ordinary people having to sacrifice what they hold dear. Such solutions will work. But, I will argue, they threaten the interests of the corporations, and they fly in the face of the economic policies of almost every mainstream political party in the world. So the corporations and mainstream parties are wedded, as we shall see, to strategies which will not solve the problem. Because of this, the only way to break the impasse is to mobilise ordinary people. This cannot be done by a movement that focuses on what people will lose to save the climate.

There is another reason for arguing against the idea of sacrifice as the solution. In the next few years each heat wave, each great tropical storm, each flood and each famine will present activists with a choice. They can side with the poor and ordinary people who will be most of the victims. They can argue against further sacrifice by the victims, and for general solutions to halt global warming. If they do so, there is a chance to mobilise the victims, and the people all over the world who feel for them, to halt climate change.

Or the climate activists can talk about general sacrifice. If they do that, they will find themselves without the support of ordinary people. They will also find themselves appealing to governments to force people to sacrifice. This will be because they will see the "greed" and self-interest of ordinary people as the problem, and top-down action as the only solution.

If and when abrupt climate change really takes off, the question of sacrifice will become acute. It will then be obvious to everyone that very large changes will have to be made quickly. In that situation, the corporations and the rich will argue that the only way forward is by forcing great personal sacrifices from the mass of people.

By the time we reach that situation, however, there will without doubt be a global movement of people who have been trying to halt the worst ravages of climate change. If the idea of sacrifice as the solution is dominant in the heads of activists in those movements, they will find themselves lining up with the corporations, the rich and the mainstream parties. They will then support sacrifice by the majority and become the enforcers of global disaster.

If the idea of social justice and sharing is dominant in the heads of activists in that movement, they will be able to mobilise the people of the world to survive, share, help each other and make the established powers of the world sacrifice.

So the argument about sacrifice bears on what solutions are possible, who can solve the problem, what sort of movement to build, and how to react to climate disasters as they happen. At root, climate activists face a choice. We can look to the ruling class, the wealthy and the powerful, for the solution. Or we can look to ordinary people to find ways to force change at the top.

The choice must be determined by what we need to do to save the climate. The decisive reason is not that poor people are nicer, or that I believe in equality and social justice. It is that the problem is so acute that it can only be solved by social solutions.

The Scale of the Problem

SOLUTIONS THAT COULD WORK NOW

CHAPTER 4

Emergency Measures

PART Two is about solutions. It is full of technological details, because I want to *show* you how solutions could work. But the detail is there for a reason. The next few chapters make five basic and important arguments.

The first key argument is that we already have the technology to stop global warming. Given the political will, we can do it swiftly across the globe.

The second argument is that putting these technologies in place will require government intervention on a global scale, including public works, legal regulation and massive spending.

The third argument is that market solutions won't solve the problem. The central failing in market solutions to climate change is that companies have to make a profit. Governments can do what the market cannot, because governments don't have to make a profit.

The fourth argument is that government solutions do not necessarily require sacrifice on the part of ordinary people. In the opening chapter I made the point that solutions to climate change which "cost too much" in fact mean many more jobs. That's what the cost is—paying people to work. Things that "cost too much" don't work in market terms. But the government is able to pay for things that cost too much, like solar power and new rail networks. So solutions introduced through government intervention don't have to mean sacrifice, any more than public roads and public hospitals do. All the way through Part Two I am also looking for the opposite of sacrifice—for the changes that will enrich people's lives while stopping climate change.

The fifth argument is that personal consumer choices cannot solve the problem. With personal solutions people are urged to examine their "carbon footprint", and then cut the emissions they are responsible for. So they can stop taking flights, sell the car, get a bicycle, change the light bulbs, buy a micro-wing turbine, buy locally produced food, and so on.

The great strength of personal choices is that they bear witness. When people change their consumption, they provoke hundreds of conversations with friends, family and co-workers. But personal choice strategies have several weaknesses. For now I will only mention two, which run through many examples in this chapter. First, most individual choices make sense only for the richer people in the richer countries. Second, personal choices can persuade some people to stop doing some things. But only massive government investment can make it possible for most people to make those choices.

World War Two

The easiest way to show how government action could work to fight climate change is to explain what happened during World War Two. Back then every major power reorganised their whole economy in order to kill as many people as possible. We need to do the same now, but in order to save as many lives as possible.

Nearly 70 years have passed since World War Two, and most people under 50 now find it hard to visualise what a government does when it really wants to solve a problem. So let's look closely at what happened then. I shall concentrate on the US example. This is not because it is special. Nazi Germany, Stalin's Russia and Conservative Britain mobilised their economies in much the same way. But I have selected the US because it was the stereotypical "free market" economy. If it could do what it did in the Second World War, anyone can do it now.

The American example is important for three reasons. The first is to show what a government does when it really means it. The second is to learn what kind of changes it made in the economy that we could learn from to cut carbon dioxide emissions. The third is to show the crucial difference between then and now. Put simply, back then business, politicians and ordinary Americans all wanted to win the war. Now business and politicians don't want to stop climate change in the same way.

America rearms

In 1939 world war began between Germany, Italy, Japan, China, Britain and France. It was quickly clear that machines would win the war. Tanks beat infantry. Ships supplied industry and the army. Tanks and ships were worthless without control of the air. Control

Solutions That Could Work Now

of the air went to the armed forces with the most and fastest planes and the most and best anti-aircraft guns. Industry built tanks, ships, planes and guns. The countries with the most productive industries would win the war.

So every country in the war strained its industry to the limits. Then, on December 7 1941, the Japanese Air Force bombed Pearl Harbor in Hawaii and the US went to war with Japan and Germany. Before December was out, President Franklin D Roosevelt sent a message to Congress listing what he wanted from American industry. The list included 60,000 planes, 45,000 tanks, 20,000 anti-aircraft guns, and eight million deadweight tons of shipping. The cost of the shopping list was estimated at $50 to $55 billion dollars, roughly the same as the estimated GNP of the USA for 1941.[1]

Roosevelt got what he asked for, because all the major competing groups in politics were pulling for him. The people who ran the big corporations believed, correctly, that if the US won the war then American business would dominate the world. So the big corporations supplied hundreds of "dollar a year" men who worked for the government for an annual wage of $1. Charley Wilson, the head of General Motors, went to work for Roosevelt's War Production Board.

The unions, liberals, African-Americans and Jews mostly saw the war as a struggle to defeat fascism. The unions and the liberal Democrats sent Sidney Hillman of the Garment Workers Union and hundreds more to work alongside Charles Wilson.

The new War Production Board did several things. The board issued "conversion" orders that told corporations to make new products. The big car corporations switched to defence orders. They made weapons, ammunition and jeeps. And by the end of the world the car plants had also made "over four million engines, 2.6 million military trucks, over 50 thousand tanks and 27 thousand complete aircraft". At Ford's new Willow Rouge aircraft plant 30 miles north of Detroit 43,000 workers built 8,685 B-24 bombers. In 1941 it took an average of 245 days for a shipyard to make the new "Liberty Ship". By late 1943 the average was down to 39 days, and Kaiser Aluminum could do it in 19.[2]

This is the first thing we can learn for stopping climate change. If the political will is there, governments can tell companies what to make. They could, for instance, tell General Motors now to stop making cars and start making buses and wind turbines.

The War Production Board also issued orders for "curtailment". These told companies to reduce the amounts of goods they were making or stop making some goods altogether. For instance, they banned almost all new house building, so construction companies were free to build new factories and army bases instead.

That's the second thing we can learn for stopping climate change. For example, at the moment cement manufacture alone causes 6 percent of global emissions. Given the political will, governments could simply tell cement companies to cut production by three quarters.

The War Production Board also controlled the flow of all materials. They told each steel company how much steel of what kind to make and which factory to send it to. Then they told the factory what to make out of the steel. They did this for all important materials, not just steel.

That's the third thing we can learn. In many industries now the least efficient plants use three times the energy of the most efficient. Governments, if they wanted to, could simply direct materials to the most efficient plants.

Everyone involved in the War Production Board experienced it as chaos from beginning to end. The chaos was worst in the first year. But somehow it all worked, and it worked *fast*. The secret was the extent of support. Amid all the chaos business executives, unions and workers were all trying to make the War Production Board work. The board were constantly drawing up the contracts and paperwork after the work had already been done. Real planning works like that. No one was waiting until the subcontractors had been selected.

This is the fourth thing to learn. We will only really tackle climate change if both the people at the top and the people on the shop floor are determined to do it.

The War Production Board created jobs

The next thing to learn is that, even though the War Production Board measures cost a colossal amount of money, they created even more jobs.

In the US in the 1940s the government built factories. Between June 1940 and December 1944 the government provided more than $16 billion of the $26 billion invested in new manufacturing plant and equipment. In less than four years that new investment increased the total value of all factories and plant by almost two thirds.[3] This had to be paid for. In 1940 only 7 percent of Americans paid any income tax. Four years later 64 percent paid income tax.

Solutions That Could Work Now

Corporate taxes also went up from 24 percent to 40 percent, and the "excess profits tax" was 95 percent.[4]

The government also made bank lending easier, sold the public "Liberty Bonds" to pay for the war and simply printed money. The national debt grew from $43 billion in June 1940 to $270 billion in June 1946. And everyone saw that the world did not come to an end with a budget deficit. Instead by the end of World War Two the USA was the world's leading industrial power.

Moreover, the armaments boom of World War Two pulled the whole world out of the Great Depression. In Britain, the USA, Russia, Japan and Germany almost everyone who wanted a job had one. Women were pulled out of the home and put to work in factories. These new workers, men and women, spent their new wages on what other workers made. Many more people then got jobs making those things. They too spent money and made more jobs, but at enormous human cost—roughly 40 million people died in the global war.

The big corporations had backed the War Production Board, and they did well out of it. Even with a doubling of the level of corporation tax, they still more than doubled profits after tax (a 105 percent rise). The big corporations got a bigger slice of that than small business. Workers did well too, but only got half the rise corporations did. Average real wages in manufacturing went from $23.64 to $36.16 a week. Average family earnings after tax went up 47 percent in five years.[5]

What this means is that stopping climate change does not have to mean sacrifice for ordinary people. It will mean jobs, hundreds of millions of them around the world, building wind turbines, installing solar roofs, insulating houses and the rest.

Rationing

The Second World War holds another lesson for climate change policy. Something happened then that politicians tell us Americans won't stand for now—rationing.

In many countries rationing was caused by food shortages. In America, far from the war and full of food, the main problem was that jobs were so plentiful that people left the worst ones. The slaughterhouses and tanneries couldn't find enough workers. So there was a shortage of leather, and the government had to ration shoes for civilians. Low paid lumberjacks and timber workers went into the factories, and the government had to ration timber and paper. Black and white sharecroppers left the southern cotton fields, but the military still needed cotton uniforms. So the government rationed civilian clothes.[6]

These days they tell us Americans love their big cars too much to ever give them up. Back then the tanks, ships and planes needed the oil. So civilians were allowed three gallons of gas per family per week. They accepted that, because they believed in what their government was fighting for. The government encouraged rail travel, and it went up fourfold, from 9 percent of travel miles in 1938 to 35 percent in 1944.[7]

Britain, Germany, Japan and the Soviet Union had rationing too. Clothes, gas, shoes and other consumer goods were rationed. More important, so was food. Every person or family got coupons or ration cards that entitled them to basic foods at the shops every week.

The rich could still afford more luxury goods, but the basic food-stuffs were shared enough for people to survive. In Britain under rationing ordinary people ate better than they ever had before. They had more money too, because everyone who wanted a job could have one. The communist Ernie Roberts, for instance, was fired for union agitation from six different jobs in six different factories in Coventry in one year. Ernie wasn't bothered. He always walked into another job the next day.[8]

The result of rationing and full employment in Britain was that during the war years infant mortality fell by 10 percent and maternal mortality by 40 percent. By 1944 the average school child was ¼ to ½ inch taller and 1.5 to 2 pounds heavier than in 1939.[9]

The lesson for climate change is that that people will accept limits on many items of consumption if they believe it is for a good cause and their lives are improving.

The money's there

All this happened in America, the proverbial land of unfettered business power. It was not just that the War Production Board could intervene from time to time and nudge business in a certain direction. They controlled what all business did, week by week, for the entire duration of the war. And they radically changed most of the American economy at the beginning of the war in six months flat.

The pace of change after America entered the war was much faster than the buildup for America's two invasions of Iraq. It was

Solutions That Could Work Now

of an order of magnitude faster than any proposals now to do anything about global warming. The American government and corporations did it so quickly because they had to win the war. The equivalent on a global scale now would be to spend next year an extra amount equal to the gross national product of the world last year—about 50 trillion dollars.

With that kind of money, planning and commitment, we could halt global warming in the time it took America to win the Second World War—three years and nine months. In fact, we could end global warming just by spending an amount equal to the current US gross national product—about $13 trillion.

So the money's not the problem, but the commitment and follow through are. The governments of the US, Britain, Germany and Japan all really wanted to win the war. But a War on Climate Change is not a war for global economic domination. So the politicians and business leaders are not desperate to fight. On the contrary, the necessary changes in the global economy threaten much of what they hold dear.

Still, the example of World War Two shows us what can be done, and the scale of what must be done. It shows us that the problem now is not the money or the technology. It's that the people in charge are not seriously committed to stopping global warming.

What will work now

The next four chapters are about solutions for climate change that can work *now*. I restrict myself to what we can do now with the technology we already have, rather than a discussion of solutions that may one day be possible with enough research and investment. There are several reasons for this.

First, I want you to believe me. It's easy to paint rosy scenarios for what will happen once we bring down the price of solar cells and hydrogen cars. But there is always a doubt in my mind when I read such scenarios.

Second, many environmentalists have a deep distrust of "technological fixes". This comes from bitter experience. Companies often refuse to stop polluting, but promise to develop a technology at some point in the future to clean up the damage they are doing now. I want the discussion of realistic, possible solutions to convince environmentalists. So I'm not going to promise pie in the sky. I'm going to pass you a piece of a pie like the ones people have been eating for years.

The third reason is that George Bush and the oil companies are now calling for research in order to avoid action. They can no longer get away by simply denying there is a problem. So their second line of defence is that we have to spend a lot of money on research to find new ways to tackle the problem. We won't have compulsory cuts in emissions until we have finished doing the research, they say.

This is hypocrisy. But this "research" approach also misunderstands the nature of technological progress. Scientific progress in basic understanding does come from research. That is why we need a great deal of research right now on how the climate works. Technological progress, however, comes from mass production, not research. In the process of developing mass production, thousands of engineers devote themselves full time to thinking about how to save money. It is simply part of the job. The result is that in a wide range of industries each doubling of production has meant energy savings of 10 to 20 percent. This is also why computers double in power and halve in price roughly every 18 months.

Of course there are "mature technologies" where most of the possible savings have already been achieved. But no one thinks wind turbines, solar cells, tidal power and wave power are mature technologies. That's why we need mass production now.

Not waiting for the research will also force corporations to act. For example, for the last 30 years American car companies have been saying that they are developing electric cars and hydrogen cars. They regularly trot out one or another prototype car to show the reporters at motor shows. The purpose is to show they have green credentials and to dodge pressure for controlling the gas mileage of the cars they actually make. But they never actually manufacture the electric and hydrogen cars.[10] If cars were banned in cities and replaced with public transport, we would find out pretty damn quick if the car companies can actually make electric and hydrogen cars. And if they can, they will be rolling off the assembly line in millions.

The fourth reason for sticking to what can be done now is that we have to act now.

Cutting emissions

Any discussion of ways to cut CO_2 emissions has to start with where most emissions come from. There are two ways of thinking about the sources of emissions. One is to look at how energy is made. In these terms almost all carbon dioxide emissions come from burning

Solutions That Could Work Now

fossil fuels. Coal creates 40 percent of global emissions, oil 40 percent and natural gas 20 percent. If you look at the problem this way, the solution is to replace these with energy sources that use other fuels. Right now that means mainly wind and solar power, backed up by wave and tidal power. Chapter 5 lays out how to do this.

The second way is to look at where energy ends up—"end users". Here the three main solutions are energy efficiency, reducing energy use, and switching fuels. Chapters 6 to 8 show how we can do this.

However, many of the solutions often proposed either don't work yet, or will never work, or have dreadful consequences. These include hydrogen cells, big dams, carbon capture and storage, biofuels, peak oil and nuclear power. What all these methods have in common is that they preserve the power of great corporations. If they worked, it would be wonderful. But they don't. Chapter 9 explains why.

Running through the next few chapters is a constant dilemma. I can easily draw up a wish list of ways to cut emissions drastically. But this means nothing unless very large numbers of people are prepared to try to make it happen. So my prescriptions for a better world can easily start to sound like fantasy. To make the fight for global solutions concrete, to make it imaginable, we have to fight in the here and now for measures that point towards that global solution. These have to be things that people can imagine fighting for, and things that can be won. So at the end of each of the next four chapters I suggest possible immediate campaigns.

Chapter 5

Clean Electricity

THIS chapter is about clean electricity. The main clean energy solutions that will work now are wind power, solar PV cells and concentrated solar power. But they have to be backed up by wave power and tidal power.[11] In a moment I will explain the technical details. But first, all clean energy solutions run head-on into a major economic and political problem. An extensive network of mostly coal and gas fired electrical power plants is already in place. These plants are connected by regional, national and international grids. These grids are connected by millions of miles of power cables.

Gas and coal fired power plants are extremely expensive to build, and they last 30 to 50 years. This means that the power companies have already invested enormous sums in their plants. If these plants are replaced by clean energy, all that investment will be "stranded". No one will need the old power plants. This would threaten bankruptcy for the power companies. Moreover, the power companies finance much of their investment with loans from the major banks, which would then also face huge losses. The power companies are a significant political force in every country, and the banks are a central power.

The way that the power companies and banks phrase this in practice is to say that "renewable" clean energy is all right on a reasonable scale, by which they mean supplying 20 percent or 30 percent of the grid. Beyond that level, they say, clean energy runs against what they call "intermittency".

Intermittency is this: wind does not blow all day every day; wind turbines work much better when the wind is really strong; solar power does not work at night, and it doesn't work as well when it's cloudy; so the supply of clean energy is unreliable.

The difficulty is that engineers have found no effective way of storing electricity. There are batteries, but they only work on a small scale. Think of how big a 12 volt car battery is, and now imagine the size of battery needed to store electricity for a city of two million people. So electricity from power plants of any kind has to be uploaded onto the grid immediately and sent out to the users, and what is not used is lost.

This creates real problems for any electricity grid in meeting "peak demand". Peak demand in hot countries is in the middle of the day, when air conditioners are running and industry is working. This is less of a problem because that's when the sun is shining, and air conditioners run most on the sunniest days. In cold countries peak demand is in the evening, when people are at home and running their heating.

To meet peak demand electricity companies need reliable power sources. They say that means they will have to use new clean energy *and* keep their old coal and gas fired plants as backup. That will mean considerable expense. And this, in turn, creates another problem for the electricity companies. It is all very well, they say, to claim that wind farms and solar power *can* provide large amounts of power. But in practice, most of the time they can't reach peak power demands. So, they argue, many wind and solar power facilities will have to be built to make sure of having enough electricity when it is needed. One wind farm may not be expensive, but the four wind farms needed to back it up on a bad day are.

So some of the objections the electricity companies have to more clean energy concern their own investments and losses on future profits. However, the problem of intermittency is real enough—but here too the companies avoid and deny other solutions. The real solutions are not local. The only way round intermittency is to combine different kinds of clean energy from many different places in very large grids. Let me explain.

Over a large enough area—Britain and Ireland, for instance—the wind is always blowing hard somewhere. Over even larger areas the sun is always shining somewhere. And as we shall see, concentrated solar power plants in hot arid areas can provide enormous amounts of electricity that can be sent across very long distances. So concentrated plants in the Middle East and North Africa can supply much of the needs to Europe, for instance. In North America the deserts of the US south west and northern Mexico can do the same. In China there are the arid deserts and plateaus of the west and Tibet.

With these economies of scale the problem of intermittency can be solved. But it would require political will, international grids and very large public investment. That too would threaten the profits of the power companies and their present political clout. This is the central reason why personal choices and market incentives will not produce enough clean energy. It is also why the clean energy solutions that follow are not alternatives to each other. To overcome intermittency we will need all of them, each compensating for the weaknesses of the others.

Wind power

Let me now turn to the realistic sources of alternative fuels. I will start with wind power, the most developed and cheapest form of clean energy.[12] My key point in this section is that wind power requires enormous investment in tens of thousands of large wind farms.

Wind, for all its virtues, cannot be a solution for individuals. Many people like the romance of their own micro wind turbine, with its echoes of old windmills. But wind power, to make any economic sense at all, has to be built in giant wind farms in windy places in the country and offshore. For many environmentalists, this reality is deeply counter-intuitive—a clean benevolent energy source should not look like a complex of monster machines. This apparent contradiction makes it harder to see that wind power is a collective social solution, not a personal choice.

Modern wind turbines are based on the same principles as the old windmills. But they look quite different, like giant propellers with three narrow blades. As the wind rises, the curved propeller blades begin to turn in the air, turning a dynamo in a round metal box housed in the centre of the blades, and that makes electricity.

The technology is simple and easy to maintain. Once the wind turbines are put in place, they mostly run themselves. They don't need fuel. This makes them cheap over 30 years. But it also means that almost all the investment has to be found at the start. So only big corporations or governments can build large wind farms.

The reason why wind turbines have to be big and in windy places is in the arithmetic. The amount of power produced is the cube of the wind speed and the square of the length of the blades. That means that if you double the wind speed and double the length of the blade, you get 32 times as much electricity. (The maths are in the footnote).[13]

The mathematics of wind turbines also explains why it's not a good idea to get yourself a home micro turbine, even though they look nice. The environmental writer George Monbiot describes the problems:

> At an average wind speed of 4 metres per second, a large micro turbine (1.75 metres in diameter is about as big a device as you would wish to attach to your home) will produce something like 5 percent of the electricity used by an average household. The most likely contribution micro wind will make to our energy problem is to infuriate everyone. It will annoy the people who have been fooled

by the claims of some of the companies selling them (that they will supply half or even more of their annual electricity needs). It will enrage the people who discover that their turbines have caused serious structural damage to their homes. It will turn mild-mannered suffering from the noise of a yawing and stalling windmill [due to urban wind turbulence] into axe-murderers. If you wished to destroy people's enthusiasm for renewable energy, it is hard to think of a better method.[14]

Wind turbines make no sense in cities and suburbs. Big turbines require strong wide concrete or rock foundations sunk into a wide area of earth. They also work best in a wind that comes consistently from one direction. City buildings simply block the wind for short turbines. For a taller turbine, city buildings still create wind turbulence that shakes and eventually wrecks the turbine. By the same token, wind turbines don't work well in a valley, which is where most cities are located. And they need a strong and steady wind. In a light wind the blades on a big turbine won't even turn. In practice, steady winds can usually be found on hills and ridges or offshore, though some deserts, tundras and high plateaus work well too.

In Europe and North America this means that most wind farms are in beautiful mountainous and coastal areas. This has provoked local battles in several areas. For instance, there have been campaigns against proposed wind farms in the Lake District of Britain and in Nantucket Sound off Cape Cod in the US.[15]

These campaigns and battles may seem like local "not in my backyard" squabbles. In fact the arguments matter a great deal. On Cape Cod, for instance, the Cape Wind corporation, the unions and the national environmental organisations stood solid for building the wind farm. Scions of several great families, including the Duponts, the Mellons, the Rockefellers and Edward Kennedy, stood staunchly against a project that was going to spoil their view while yachting from their summer homes. The rich families were able to knobble local and state politicians, and to tie up Cape Wind's plans in the courts for seven years to date. Cape Wind is run by a very determined boss with deep pockets, and he is still trying. But the result of these delays is that other companies have seen what happened to Cape Wind. No other company has tried to build an offshore wind farm anywhere on the Atlantic coast of the United States. What seems like a local dispute about the "environment" has in practice been a battle about whether wind power in the US will make a difference to climate change.

In fact, it is important to realise that this is not a debate about the "environment"—it's about how things look.[16] Yet climate change is not about how things *look*. You can't see carbon dioxide. Many people think of environmental problems as being mainly about appearance—about billboards and the beauty of the wilderness. But global warming is not that kind of problem. It is more like war and famine than it is like litter. The question is not whether the solution is beautiful, but whether it will save the Earth.[17]

In short, wind turbines work. They have to be put in windy places, but the world is full of those. In Britain one estimate is that the wind energy available onshore is three times greater than all Britain's expected electricity needs in 2030. On top of that, offshore wind farms could provide six times all of Britain's electricity needs. On a global level, a 2005 study by Christina Archer and Marc Jacobsen of Stanford University estimates that wind farms on land and near shore could provide 100 percent of global energy demand and seven times the current global demand for electricity. This is a conservative estimate—Archer and Jacobsen only looked at the windiest regions where wind power could be produced most cheaply.[18]

Solar power

Solar power uses sunlight to make energy. The two key forms are photovoltaic cells and concentrated power.[19] It is a tried and tested source of clean energy. But even more than wind, solar power in all its forms requires government intervention to work. This may seem surprising, because we are all familiar with solar panels on private houses. However, almost all of these have been supported by some form of government subsidy. This is because solar power is simply more expensive than wind power.

Photovoltaic (PV) cells are what people usually mean when they say "solar power". PV cells are thin sheets of silicon that turn sunlight into electric current. The current can be plugged straight into the existing electricity system. The main method for using PV cells is to cover rooftops with arrays of cells that look like flat roofing panels. Solar panels can also be put on top of old roofs. But they are cheaper on new buildings, because the contractor can save the cost of roof tiles. In hot sunny countries solar panels are put on all roofs. In cold northern countries only the south facing roof is covered, because it gets more sun. In the Southern Hemisphere it's vice versa.

Solutions That Could Work Now

PV cells save a lot of money because you don't have to move the electricity. With conventional electricity, half the cost is in making the electricity at the power station. The other half is the cost of the power lines to houses and the electricity that is lost because it leaks out of the cables.[20] That half of the cost is eliminated by using PV panels right on the roof.

But even with that saving, PV cells are currently much more expensive than wind farm energy, not to mention coal, oil and gas. It is difficult to tell how much more expensive, because all the figures for the prices of solar energy are suspect. The enemies of clean energy exaggerate the costs. But most writers on solar energy either support it for environmental reasons or work in the industry. Almost all of them assume they have to make a case for solar power being competitive in the market, either now or soon. But it is not possible to make the case for solar power in terms of strict profit and loss. Since they can't think outside the market, they are forced to waffle. So they massage the figures, or compare costs to imagined future oil prices.

Still, the cost of PV cell electricity will come down with mass production. The real cost of conventional electricity in the US in 1900 was 140 cents a kilowatt hour (in 1990 prices). By 1925 it was 40 cents, and in 1980 it was 6 cents.[21] We can expect the costs of PV cells to drop in the same way eventually. But I have said that I will stick to solutions that work now. Moreover, to make those savings from mass production, someone has to do mass buying while it is still expensive. The only solution is government subsidy. This has been tried in Japan and Germany and it has worked. Those two countries currently account for 69 percent of the global market for PV cells.[22] And neither Japan nor Germany has a lot of sunlight.

There are four different possible ways of subsidising either wind power or solar power. One is a law requiring the electricity grid to buy a certain percentage of its power from clean energy. This method favours wind farms, which are much cheaper than PV cells. For instance, as a result of a favourable renewable energy law passed under Governor George W Bush, by 2007 the state of Texas, with half the population of the United Kingdom, had more than two and a half times the wind energy installed. One wind farm in Horse Hollow, Texas, accounted for almost 40 percent of the total wind capacity in the UK.[23]

The second form of subsidy is used in Japan—a grant to homeowners. This started because the Japanese government was nervous about the security of Indonesian and Middle Eastern oil supplies. So it began the "70,000 Solar Roofs" campaign in 1996. The government paid half the cost for anyone wanting to install PV cells. Sales grew 50 percent a year from 1995 to 2005. In those ten years mass production meant the price fell by half. By 2004 it had fallen so far that the government was able to lower the subsidy to 7 percent.[24]

In Germany the driving force behind solar power was the Green Party. They wanted to stop climate change. The Green Party were junior partners in a coalition government with the Social Democrats. That government launched a campaign for 100,000 solar roofs.

Germany used the third possible method of subsidy, the "feed-in tariff". This relies on the fact that the PV electricity from the roof can be fed straight back into the main electricity grid. The PV cells are making electricity all day, but the household will only be using all of this power some of the time. So the German government worked out a standard rate the electricity company had to pay for renewable energy. This price was set high enough that the person or company who installed renewable energy could recoup their investment in ten years.

By 2006 Japan and Germany between them accounted for just over two thirds of the global market for PV. In third place was the United States. In America the federal government under Clinton and Bush did nothing. Instead subsidies came from state and city governments. As of 2006, 21 states used Japanese-style subsidies for solar power.[25] Tucson, Arizona, and Sacramento, California, offered up to $3,500 per installed kilowatt of production. Los Angeles offered the same, and aimed for 100,000 rooftops by 2010. In 2006 the state of California, under Arnold Schwarzenegger, passed a law aiming for *one million* solar roofs.[26]

There is a fourth possible kind of subsidy. The government doesn't pay half the cost, like in Japan. It pays all the cost.

On the face of it, this proposal may sound mad. Visionary. Utopian. Socialist. If you feel that way, I have one word to say to you—"sewers".

In the 19th century the people who ran the new industrial cities of the world faced a massive environmental problem. They had to get clean water to all the city dwellers and businesses. And they had to get the dirty water, the shit and toxins, away. As the century went on, moreover, they began to understand that germs carried in water caused diseases. So they built sewers, and systems for filtering the water, and systems for delivering clean water. Those sewer systems were very expensive public works. But they were necessary to solve a massive problem of pollution and keep people healthy.

There are other words I could say. "Roads", for instance. "Bridges". "Schools". "Hospitals".

Solutions That Could Work Now

Concentrated solar power

PV cells are one of the two main forms of solar power. The other is "concentrated" solar power. Concentrated power is particularly suited to large-scale investment, and to providing electricity over very long distances.

Solar power will always be relatively inefficient in Germany, Japan and Britain. Yet it can be very productive in the hot, permanently sunny places in the world. But "concentrated solar energy" works even better than PV cells. With concentrated solar energy giant mirrors are used to direct and concentrate the light of the sun. It's like the girl scout trick of starting a fire in a piece of rotten wood with a hand mirror, but on a giant scale.

The light from the mirrors can be concentrated on PV panels. But it produces far more energy if it shines on water, oil or liquid salt in a pipe. This produces temperatures of up to 400°C that force steam along the pipe and through a generator. The mirrors are built so they move during the day, tracking the sun and constantly concentrating as much light as possible. Concentrated solar power farms have been used in California since the 1980s, and are now found in Spain, Italy, Morocco, India, Mexico and several other countries.[27]

Moreover, long range cables can carry direct current from deserts all over the world to the rich countries. According to George Monbiot, "The International Energy Agency calculates that if solar photovoltaic panels were used to cover 50 percent of the land surface of the world's major deserts, they would produce 18 times as much energy—which means 216 times as much electricity—as the world now uses".[28]

And that's just with solar panels. But concentrated solar power also makes it possible to avoid intermittency. In the hot places of the world concentrated power facilities can heat the water, oil or salt and then save that liquid, still hot, for the night time, when it can be released to power the steam dynamos.

The obvious problem is that concentrated solar power works best in deserts a long way from the electricity user in Toronto or Berlin. A recent technological breakthrough—long range high voltage direct current cables—has solved this problem. Traditionally electricity has been sent through the grid in cables that carry alternating current (AC). But AC cables constantly leak and shed electricity, making it very expensive indeed to send electricity over long distances. Direct current, on the other hand, loses almost none of its electricity along the way. The new DC cables work with almost no loss of power over long distances. There is already a DC cable 1,700 km (1060 miles) long in Congo.[29] Longer lines appear perfectly possible.

This allows for the mixing of wind and solar power over such long distances as to largely solve the problem of intermittency. Long distance cables also offer another possibility. Engineers and scientists in the German Aerospace Agency, working in partnership with governments in the Arab world, have already developed a magnificent plan called TREC. This would use concentrated solar power plants in the deserts of North Africa and Arabia to generate electricity that could be sent all over Europe. They have published two book length reports with detailed plans for the supply of 30 percent of Europe's energy, using current technology.[30]

The reports estimate that this would require an initial investment of $75 billion. This sounds like a lot of money. But they estimate that over the long run it would avoid $250 billon of costs on conventional energy. More important, $75 billion spread over ten years is $10 a year per person in Europe, or 80 pence a week for a family of four in Britain.[31]

That would also mean jobs for the people who would otherwise lose their livelihoods with a switch to clean energy. The ordinary people in the oil countries of the Middle East have endured many wars and dictatorships because of the wealth they sit on. It would be an ugly tragedy if clean energy pushed them into destitution. Of course there will still be opportunities to use oil for fertilizers and plastics. But there would be a deep justice if the people of Saudi Arabia, Iraq and Afghanistan could become the generators of solar power for the world.

Transporting electricity on DC cables does mean building a whole new global network of electricity pylons. These would have to go to different places for supply from the current network. DC pylons are smaller and lighter, however, and therefore less expensive. The main cost is not the manufacture and installation of the cables. It is obtaining the right of way and land. This again is a task for governments.

The TREC plans for electricity from the Middle East and North Africa are ambitious. But the German Aerospace Agency has also been careful to make proposals that would be cheaper than the alternatives. A much larger project would be perfectly possible, although more expensive. The German report says the amount of energy potentially available from solar power around the Mediterranean region is "several orders of magnitude larger than the total world electricity demand".[32] That means at least several thousand times the current electricity needs of the whole planet.

The figures for long distance concentrated solar power have so far only been worked out for the region of Europe, the Middle East and North Africa. Similar plans would be as easy technically, and politically easier in Australia; in North America, using the deserts of the American south west and northern Mexico; and in China, using the sunny and windy deserts and plateaus of the far west and Tibet.

Other realistic solutions

Wind, PV cells and concentrated solar power are the best developed technologies for clean energy. But other forms of energy will be needed to balance the supply of energy to the grid at times when both wind and solar are weak. The main technologies now coming on stream are tidal power and wave power.

Tidal power uses the energy rushing in and out of tidal estuaries. A barrage is thrown across the entrance to an estuary, and the incoming and outgoing tide turns dynamos to make electricity. Wave power relies on incoming ocean waves to turn small motors in many dynamos strung in a line through the water. The great advantage of wave power is that waves are steady and reliable, so they do not suffer from the intermittency of wind, solar and tidal power.

Both these technologies look promising, although they are currently much more expensive than carbon fuels. Neither can work as personal choices—they are major government projects. They have been very little developed because of market constraints. But again that means that big government projects for wave and tidal power would create jobs.

All these options taken together can provide a mix of clean energy sources that will easily supply 80 percent or more of electricity generation. The different energy sources, in different places at different times, can combine to provide the total electricity needed. The key is long range cables and very big grids.

Such measures can overcome any problems of intermittency. It is difficult, in any case, to tell how large this problem is. One recent study estimates that paying for back-up if clean energy supplied 30 percent of the British grid would cost an extra £330 million to £920 million. This is not that much, particularly as the same study estimates the cost of production would be £9 billion anyway.[33]

So the costs for back-up are easily affordable if clean energy supplies 30 percent of the grid. Above 30 percent costs do start to escalate. As to how much they escalate, though, your guess is as good as mine. No one who could has yet done the calculations. However, a recent study by Mark Barrett of University College London does provide a convincing plan for 95 percent of UK energy needs just from renewable sources within the country.[34]

It is also true that making wind turbines and PV cells burns up a lot of energy in mining the materials, processing them in a factory, and transporting them to the site. Right now, all those activities create considerable CO_2 emissions. But much of this problem will go away in time. Building the first wind turbine or solar array burns a lot of carbon. Once the world begins to fill with alternative energy, mines, factories and transport can use electricity from wind turbines.

But the key point in any discussion of clean energy is that we don't have to produce it within the limits of the market. If it costs more, it costs more. Of course, there is no point in simply wasting money. That is why it makes sense to pay for solar panels, but it does not make sense to pay for micro wind turbines. But spending that will never make a profit makes sense. Coal, oil and gas fired put carbon dioxide into the air. Wind turbines, PV cells and concentrated power do not. They are easy to build, they are quick to build, and we can rewire the world.[35]

What do we fight for now?

At the start of this chapter I said I would look for specific concrete demands people could fight for now. Of course any real demands have to be worked out by meetings of people who represent unions, campaigns, and parties who might lead a movement to make them a reality. But let me suggest two ideas that might work in Britain and one idea for sunnier countries.

My first suggestion starts from the fact that Germany already has ten times as much wind power installed as Britain does. There is no reason Britain could not have double that, 20 times the current wind power, in five years. Another way of saying the same thing is that Britain could build ten wind farms each year the same size as the one in Horse Hollow, Texas. After all, Britain is the country in Europe with the best supply of both onshore and offshore wind. The British weather would finally be good for something.

My second suggestion starts from five facts. One—the Californian state law providing for government subsidy to fit one million rooftops with solar panels.[36] Two—Germany already has 20 times as much installed solar power as Britain.

Solutions That Could Work Now

Three—in Britain a government programme used to provide grants for people wanting to install solar panels or insulate their houses—£10,000 for PV cells and £5,000 for a wind micro-turbine. In order to ration the grants, the government decided that applications would only be accepted at one website on one day of the month. On that day the website only remained open for about 45 minutes each month, by which time it had received all the applications for which it would pay out. Most people who tried to get through to the website during those 45 minutes were unsuccessful. This was so embarrassing for the government that in May 2007 they cut the maximum grant to £2,500. This meant that only the wealthy would be able to afford solar panels. But we know there is a large pent up demand for grants for solar panels.[37]

Four—in the summer of 2007 the new British prime minister, Gordon Brown, announced plans to build three million new private and public homes. Five—the New Labour government and the Conservative opposition both go on endlessly about how much they care about climate change.

Those five facts seem to me to point to a simple demand. Britain has twice the population of California. If the Terminator can install a million solar roofs, we can do five million. Much of that can be done simply by requiring that all new homes and public buildings have solar roofs. This is much the cheapest way to install solar panels, because you save on roof tiles. The remainder of the five million solar roofs could be funded by government grants.

That's an ambitious demand—five million solar roofs in five years. But ambition inspires people, and you can't build a campaign without passion. It is also a demand that could be won. It is being done elsewhere. It's a concrete demand, easy to understand; it means jobs, and would immediately start Britain towards a clean energy economy.

Those are the obvious campaigns in Britain. In the sunny parts of the world—in places like Egypt, Mexico, India, the Sahel, Australia and the south western US—a campaign for concentrated solar power may make more sense. This should not wait upon the construction of long range cables. Instead concentrated solar power could be used to meet all the region's own energy needs. Once that was working on a large scale, the rest of the world would see what was happening. This is how it works with oil now. You don't wait for the tankers. You show you have the oil, and then the tankers will come. In countries like Egypt, Sudan and Morocco this has the possibility of transforming the economy and beginning to transform the world.

Chapter 6

Buildings

THE last chapter was about transforming electricity to eliminate CO_2 emissions. The next three chapters will be about cutting energy use in buildings, transport and industry. Again I look for solutions that work now, with the technology we already have. I show why those solutions require massive government intervention, and cannot work with market incentives and personal consumer choice alone.

There are three main ways of saving energy. One is energy efficiency: using less energy to achieve the same result. For example, you insulate leaky houses and then you can heat them to the same temperature using half the electricity.

The second way to save energy is to change energy use. For instance, cement manufacture emits large amounts of CO_2. So the government could ban the use of cement in buildings. This is akin to the way the US government required construction firms to stop building new homes during World War Two. The third way is to change the fuel used. For instance, the government can ban cars in cities and provide trains and buses run on electricity. If that electricity is generated from wind and solar power, CO_2 emissions plummet.

To drastically reduce emissions we will need all three methods together—energy efficiency, change of use and change of fuel. Most writers on energy saving, however, concentrate on energy efficiency. The reasons are political, not technological. Most corporations can imagine better energy efficiency. But changing the use of energy would be a direct challenge to the cement, construction and oil companies. Changing fuel use would challenge the oil and car industries. So most writers try to be "realistic" and stick to energy efficiency. The trouble with this realism is that it is wildly unrealistic about what must be done to avert climate catastrophe. Running through this whole book is the idea that direct challenges to corporate power are necessary to stop climate change. There is no way round this.

The end users

Before we turn to the detail of energy use, let's look at the total picture of where global carbon emissions come from.

Because man made global warming is a result of the industrial revolution, many people assume that most emissions come from industry. This is not the case. Globally, buildings and transport together are more important than industry. Moreover, CO_2 emissions are concentrated in particular areas of housing, transport and industry. This means that we don't have to change the whole of human life. Large, simple changes in a few sectors of the economy will eliminate the majority of emissions.

The last year for which we have accurate figures is 2004. Total global emissions of carbon dioxide that year were 27 billion tons of carbon dioxide. (They are now about 28 billion tons.) These figures do not include CO_2 emissions from cutting down forests, or methane emissions, which I discuss in Chapter 10.

Globally, industry produced 37 percent of all CO_2 emissions in 2004, homes and commercial buildings 32 percent, and transport 23 percent.[38] Agriculture and other small end users accounted for the remaining 8 percent. In richer countries industry is less important. In the UK in 2006 the split was buildings 37 percent, transport 34 percent and industry 29 percent. It was very similar in the USA in 2005—buildings 38 percent, transport 33 percent and industry 28 percent.[39]

The problem, however, is not in all building use, or all aspects of transport, or all industry. Here are the rough global shares of the main energy users:

Percentage of global CO_2 emissions by end users, 2004

Heating in buildings	10 percent
Lighting and appliances in buildings	8 percent
Cement making	6 percent
Petroleum refineries	6 percent
Iron and steel making	6 percent
Car travel	10 percent
Truck freight	6 percent[40]

Those seven major end users are responsible for half of all global carbon dioxide emissions. The next three chapters will concentrate on how to cut emissions in these sectors, and also look at air conditioning (3 percent) and air travel (3 percent). Of course we will also need thousands of other cuts and changes. But they will be similar in kind to the cuts and changes in the main sectors. In these three chapters I cannot, of course, cover everything, but I can show how deep energy savings are possible.

Buildings

Heating, cooling, lighting and appliances are the big energy users in buildings. In the richer countries about three fifths of building emissions come from homes, and the other two fifths from public buildings, offices, stores and businesses (not including factories).[41]

The main problem with buildings in the rich countries is heating— the single largest source of CO_2 emissions globally. My first example comes from wet, cold, leaky Britain. Here heating accounts for 85 percent of energy use in homes. I will start with new build, where cuts in emissions are easiest, before turning to buildings that already exist.

In the late 1980s architects in Germany invented the passive house. Passive houses use only the heat coming through the windows from the sun and are cooled by ventilation. The windows are positioned to face north or south, and made large or small, to allow enough heat in during the winter but not too much in the summer. The whole building is built with minimum leakage. It is insulated against the outside at every point—not only in the outside walls, but also where it touches the foundations.

Passive houses work. There are currently 4,000 in Germany, 1,000 in Austria, and a few in other countries. One study of 100 houses in Germany showed an average winter temperature of 21.4°C (70.5°F) indoors. There are considerable costs for the careful engineering, design and insulation. But the builder saves a great deal of money by not installing heating and air conditioning. The net result is that passive houses are only about 10 percent more expensive than ordinary houses.

It is also possible to use many of these innovations in windows and ventilation in new buildings that are not completely passive, and do use limited amounts of energy for heating and cooling. In this case the builder cannot save as much money, but energy savings of 80 percent are still possible.

In Scandinavia they build houses to keep out the cold, because they have to. If British buildings met the minimum standards in

Solutions That Could Work Now

the building codes of Norway and Sweden, British homes would use a quarter of the energy they now do for heating. That's a three quarters reduction in the household heating bill. Moreover, putting the insulation into a new house does not cost much extra. So it makes economic sense. However, the saving goes to the person who buys the house. The expense goes to the company building the house. So the company will not insulate new houses properly unless building regulations compel them to.

There are two possible energy efficient solutions for new houses to fight global warming. One is to require full insulation for all new houses and enforce the law. Even better would be to require all new build to be passive houses.

Energy efficient new houses can be built without tax incentives or government grants. Indeed, a government directive for all new buildings would make a much larger difference than a small number of affluent people taking advantage of tax incentives or grants. And it would create more jobs.

This is not a fundamentally new approach. Every country in the developed world has building regulations for all new houses to make them safe. These regulations have always meant that the builder has to spend more than he otherwise would. What I am suggesting here is just expanding and enforcing the building regulations to make the world safe.

Moreover, in Britain such new regulations would not actually make new homes more expensive. A large proportion of the cost of city houses is the land. Home buyers have only so much to spend on mortgages. If the cost of building the house went up, the cost of the land would go down.

Old houses in Britain

So far we have been talking about new build. But in 20 years time most of the houses people live in will be the same ones they are living in now. Of the 25 million dwellings in Britain, only 17 million have cavity walls with space for insulation. Of those, only six million have any insulation—the other cavities are empty.[42]

The real energy savings are in retro-fitting Britain's leaky homes with draught prevention, properly fitted triple glazing and good insulation under floors, in walls and in attics and roofs. This would cut energy use on heating by over half. But it is much more expensive to retro-insulate than to insulate new build, because you have to tear out walls and floors and then build them again. It costs at least £20,000 ($40,000) a house, and often much more. Even with a government grant to pay part of that cost, this is still an expense beyond the reach of most people. In many cases, it would save more energy to tear down the house and build again, but many people love their homes.

Again what is needed here is a massive government building programme to insulate or rebuild all the houses in the country. Not partial grants to encourage the affluent who are climate conscious, but systematic free insulation for everyone. This kind of improvement in insulation is needed not just in Britain, but across Europe and all the cold countries of the world. So we are talking about tens of millions of jobs for building workers.

Of course there are also carbon emissions during the building work. But as with the installation of wind and solar power, the carbon emissions can drop as the process takes off. This will only happen, however, if the rest of the economy uses clean energy. Again there is little point in energy efficiency without massive investment in wind and solar power.

Hot houses in America

My second example of possible savings in buildings comes from the United States. Many houses in America suffer from the same heating problems as in Britain. But the hotter USA also provides a particularly good example of the opposite problem—air conditioning.

My emphasis will be on public and commercial buildings. These use a lot of air conditioning because the trend has been towards offices and stores that cannot work without artificial cooling, and because people are inside them during the hottest hours of the day.[43]

Air conditioning now uses somewhere between 30 and 40 percent of the energy that goes into American buildings. That means 3 to 4 percent of the world's total carbon emissions come from American air conditioners.[44]

Air conditioning is a technology rather like the automobile. Both started out as a luxury for the few and ended as a necessity for the many. With automobiles, cities and suburbs were rebuilt to demand cars. With air conditioning, the construction industry was remodelled in its image. Public buildings, particularly office buildings, are built to require more and more air conditioning. They are tall, thin-skinned and covered in glass. In other words, they're greenhouses.

It's the fashionable look for modern buildings. Everywhere people draw the blinds in a vain attempt to keep out the heat, and they turn up the air conditioning because they can't open the windows.

Moreover, tall buildings use twice the energy of shorter buildings. Architect Sue Roaf and building physicist Fergus Nicol explain:

> The higher the building, the more it costs to run because of the increased need to raise people (lifts), goods, and services, and also, importantly, because the more exposed the building is to the elements the more it costs to heat and cool. The higher the building, the higher the wind speeds around the building, the more difficult to keep it out, and the more wind pressure on the envelope sucks heat from the structure...
>
> Lifts are very energy-expensive and costly to run, maintain and replace. Lifts alone can account for at least 5 to 15 percent of the building running costs and the higher the building, the more it costs...
>
> Many tower blocks have fallen into a very poor state of repair because their owner cannot afford their upkeep... It is often cheaper to blow up tower blocks than to repair them. "Tower blocks are only for the rich", said Michael Holmes of Arup Associates.[45]

Moreover, people don't like working in tower blocks. American office workers take an average of one day a month off due to illnesses caused by the air conditioning. Tall buildings are very vulnerable in a blizzard, a power outage, or a fire. Few tall buildings even bother with fire drills. In New York, for instance, the Fire Department does not plan to go above the tenth floor. And workers in tall glass buildings don't like the hum, the feel of the air, or the lack of privacy.

In the medium term, we will not be able to cut off all air conditioning. We are stuck with what has already been built.

The long term solution is for all new homes to be passive houses, but of a different sort. It is not necessary to find new architects to design them. I spent much of my childhood in Texas and India. Both were hot places with traditional cool houses. These had thick mud, clay and brick walls, and interior courtyards with trees and ponds. Windows and doors were arranged to catch breezes and keep everyone cool.

These houses traditionally went together with siestas during the worst heat of the day. In the winter my school in Lucknow on the north Indian plain kept normal hours. In the hot season of spring, just before the monsoon, we started school at six and were finished at noon. In the hot nights we slept on the roof under mosquito nets under the stars. This was a human way of life, fitting to the place.

When I talk about turning off the air conditioning, I do not mean ordinary people should sacrifice. I am talking here about different, not less. Different houses that fit the land. Different working hours that fit bodily and climate rhythms. People getting up to put on a sweater or take one off before they adjust the thermostat. And low-rise human office buildings with windows that open, so you can cool the building and hear the birds.

I am also talking about government regulations so that new building fits the land. The use of air conditioning in the US has exploded as people have moved south to swamps, or to semi-deserts that already had water shortages. People are moving to places that they actually do not want to live in, and making them like the places they would want to live in. The beautiful feel of the air in a tropical dawn the pleasure of a desert night and are denied to the new settlers.

In many cases, half the electricity used can even be saved just by adjusting the thermostat so it tracks the outside temperature and adjusts the inside temperature to stay cool enough, but not massively at odds with the local climate.

It needs underlining again here that cutting emissions from air conditioning is not simply a matter of reducing energy use. If air conditioners are powered by electricity produced by solar power, they use up very little carbon.

Lighting and appliances

The third area of possible saving in buildings is lighting. The traditional incandescent bulbs found in most houses mainly produce heat that is wasted, and light as a by-product. (They don't blind you when you look at them, but touch them and they burn you.) Replacing these with modern fluorescent bulbs cuts the energy use to a fifth—from a 40 watt bulb to an 8 watt bulb for the same light in my house. The 8 watt bulb costs four times as much, lasts ten times as long and saves a lot of money over its lifetime.

Lighting takes about a fifth of electricity use in the US. About half of that goes into incandescent bulbs, and half into fluorescent and other business lighting. Simply replace the incandescent bulbs and you reduce total lighting emissions by 40 percent. With different fluorescent bulbs and lighting designs, emissions can be reduced by a total of 70 percent. After that you can turn off the lights when you

Solutions That Could Work Now

leave the room, design offices and houses so natural light is allowed in, and use timers that turn off lights that are not tended.

Lighting provides a good example of why personal consumer strategies are not enough. There are now many campaigns to encourage people to buy better bulbs, educate businesses, and call for government subsidies for the cost of the 8 watt bulbs. The real solution, though, is for the government to ban the old bulbs. The Australian government has already announced that it will do this by 2010.[46]

There are also the increasing number of appliances that draw almost as much power on standby as when they are on. Computers, televisions, radios and stereos are all now built to do this. This saves a trip over to the plug to switch it off. (And we wonder why people are getting fatter?) The common strategy now is to encourage people to remember to switch off these appliances. There are also a number of ways of preserving the standby facility while using a great deal less energy. But again there is no good reason not to outlaw the manufacture and sale of appliances with such standby functions.

Changing fuels

There are two more important savings to be made in the majority of houses in the US, and in most rich countries. One involves a simple switch in heating fuel. Most houses use electricity for appliances, lighting and air conditioning. They use kerosene or natural gas for heating, and gas for cooking. Replacing kerosene and gas with electricity generated by solar or wind power avoids almost all the CO_2 emissions.

There is, however, a crucial limit on such savings. Electricity is an inefficient way of making heat. Generating electricity for heating from coal or gas in a power station and then delivering it to houses creates about three times as many emissions as heating oil burned in the house. This is why heating by electricity is now so expensive. It also means that even if two thirds (67 percent) of the grid is supplied by clean energy, there is no net saving in using electricity, and no point in switching. But if 90 percent of the grid supply comes from clean energy, then a switch to electricity cuts CO_2 emissions by 70 percent. (The arithmetic is in the footnote).[47]

A second possible saving that is often recommended comes from "combined heat and power". Electricity generation in a power station also produces a great deal of heat, which is usually vented into the air and wasted. When the power station is close to urban areas, however, the hot air can be trapped and sent to houses and buildings in underground pipes. But this requires considerable initial expense and needs government support. More important, combined heat and power do not work with wind or solar PV cells.

Campaigning now

So what can we fight for now? Very considerable energy savings are possible in houses and buildings. The largest savings, and therefore the key ones to fight for, are in heating, air conditioning and light bulbs. Here again, though, ecologically conscious green consumers will not make the decisive difference. On a global level, banning light bulbs is better than individuals buying fluorescent bulbs. A few people can afford passive houses, but everyone will have to follow building regulations.

However, I have shown what we could have if only governments would do it. And I have shown that the scale of what needs doing requires government subsidies, government public works and government regulation.

As with clean energy, it is important to think how key measures could be made into demands for real campaigns backed by organisations that can mobilise large numbers of people. As before, the demands have to be ambitious enough to mobilise people, but they also have to be for policies people believe are possible. I can think of three such demands.

The first, in Britain, is for partial or full government grants to insulate leaky old houses and fit new windows. If we can demand five million solar roofs in five years, we can demand ten million insulated homes in ten years. As we have seen, there are enough people who want government grants.

The second demand, in many countries, is to stop the construction of new tall office buildings and shopping centres. Tall buildings are uncomfortable and sick making for the people who work in them. They also waste electricity in lifts, heating and cooling. They must be heated and cooled as a whole unit, making it difficult for users to turn anything off when they leave their part of the building. And they consume large amounts of concrete and steel. Concrete and steel manufactures together account for an eighth of global emissions of CO_2.

Solutions That Could Work Now

The fight against tall new buildings does not mean an end to construction. It means people will work in offices and shop in stores that are built on a human scale. In the end the fight against tall buildings is one that must result in national laws. But it can start as many local campaigns, building by building. These can be community campaigns, particularly against giant air-conditioned shopping malls like Wal-Mart. And many office workers, and especially public sector workers, have unions. Those unions can also fight, building by building, for workplaces they can live in.

The third obvious fight is in some ways the easiest. We have to regulate light bulbs, locally and nationally.

Each of these fights has an obvious constituency to mobilise, and an obvious way to pursue the campaign. But we also have to fight for new building regulations that enforce passive and low energy houses. Although I suspect this would be more difficult to mobilise for, it is an easy thing to put into law.

CHAPTER 7

Transport

L ET me now turn from buildings to possible ways to cut CO_2 emissions from transport. As with clean energy and housing, my argument here is that it will take government action to make the deep cuts that are necessary.

Transport is responsible for about 23 percent of global carbon, 31 percent in America, 23 percent in Europe and 12 percent in Asia.[48] Here are roughly accurate figures for the share of global transport emissions:

Global share of total emissions of CO_2 from transport

Cars	45 percent
Trucks	25 percent
Planes	12 percent
Shipping	10 percent
Buses	6 percent
Rail	1 percent
2-wheelers	1 percent[49]

In other words, 82 percent of global CO_2 transport emissions come from cars, planes and trucks. Cars are the most important, but air travel is growing fastest.

Cars and Public Transport

There are two basic problems with cars. They burn oil and people drive alone.

Globally, city journeys account for almost 90 percent of CO_2 emissions from cars. So let's start with cities. Stand on the pavement during the rush hour in any major city or suburb in the rich world and watch the cars go by. You see an endless stream of cars and lonely drivers. That's the biggest problem. The car is big and heavy enough to take five or six people. But much of the time it only carries one.

Solutions That Could Work Now

The answer is public transport. Buses and trains emit much less carbon dioxide per passenger mile. The exact numbers are hard to calculate. It depends on the kind of bus, the speed of the train, and the number of passengers in the car or the bus. Yet it is possible to work out some rough numbers, using studies from Canada, Britain and Poland, and one study covering much of the developing world. (The numbers are in the footnote.)[50] The figures vary for each country, but in general a switch from cars to buses cuts emissions by about 70 percent, and a switch to trains cuts 80 percent. The figures are better if the buses and trains run fuller than they do now.

This is important in the rich countries, and particularly in the United States. But public transport matters even more because of the poor countries. People in those countries are travelling more than they used to, and they will be travelling far more in the future. That will mean one of two things—a massive increase in public transport or an explosion of private car ownership. Plan B will boil the earth. Yet it is also the case that public transport will only dominate in the poor countries if it dominates in the rich countries. That's how the world is. People in Calcutta want to do what they see people in Hollywood doing.

Consumer choice and market incentives

It's worth repeating—city travel accounts for almost 90 percent of CO_2 from cars. The solution to car emissions is to ban cars in cities and provide first rate public transport—buses, trains, underground railways, trams and bicycles. This would not mean sacrifice. Public transport without cars would not be *less*; it would be *different*. And that difference would be *better*.

Many people now prefer cars because of the current state of public transport. It's infrequent, late, slow, crowded, dirty and sometimes non-existent. Moreover, today public transport often costs more than cars. For all the journeys I make in Britain, a train ticket costs more than going by car. The government subsidises the roads, and the private train companies are squeezing out profits by putting up the ticket price.

Many people also love the freedom of the car. You can go where you want. You can put on your own music and hang your arm out of the window. For many people, it is the one place in the world where they are the master. And certainly we are told that cars are convenient, give us freedom and make us sexy. The last is not true, as driving a car everywhere makes you fat. Nor are urban rush hours oases of freedom—they are stress cookers.

So we need buses and trains that come every three or five minutes, run on time, are clean and not too crowded, and run all night.

The system also has to provide the flexibility of cars. At the moment trains and buses run into cities, but not across the suburbs. What people need is a network of connections, and transfer tickets that make connections free. They also need a choice between buses that stop every 200 metres and express services that zip along.

None of this will be really attractive, though, if half or even a quarter of people still drive their cars alone to work. The roads will still be crowded, and traffic will still creep along in the rush hour. Both bus and car passengers will be trapped. However, if everyone went by bus, journeys would be swift and there would be no traffic jams.

To be appealing, any public transport service also has to be very cheap or free. Making it free is a considerable saving in travel time and salaries. No one needs to sell, take or inspect the tickets. Buses move more quickly, with fewer staff. This can be paid for out of general taxation, in the same way that schools and medical care already are in Britain.

Moreover, banning cars can also change the nature of city life. This will make promoting buses and banning cars not a sacrifice but a joy. The benefits here have nothing to do with CO_2, but they still matter. The most important change is what begins to happen where the cars used to be. The buses and trains will have enough room on a quarter of the streets. On the other three quarters of streets there will be no vehicles at all. Children will be protected, and have far more space to play. With parked cars removed the streets will be twice as wide. Trees, bushes, grass and flowers could turn the old streets into countryside.

Cycling would also be safer and easier. Some side streets could provide cycle lanes, where children and adults could ride freely and safely, to the benefit of their health. Disabled people and the frail elderly, who are now excluded by most public transport, could have access to those lanes in small electric vehicles.

There would also be fewer disabled people. According to various estimates, cars kill between half and million and a million people a year globally, and permanently disable about 800,000.[51] And there would be massive increases in air quality, and falls in disabling and frightening chronic lung ailments.

Public transport for everyone is not a sacrifice. It's a better world *and* one with less CO_2—something worth fighting for.

Solutions That Could Work Now

Personal and market solutions

Public transport, however, can only deliver serious cuts in CO_2 *without* sacrifice *with* major direct government action. Personal choices and market incentives won't work in the same way.

Even a major campaign to urge people to use public transport runs into the problems of the actual state of the service. It's not good enough just to promote public transport. It actually has to be better, and more available, than it is now.

Market and financial incentives are also intrinsically flawed. There are two possible kinds of incentive—either penalise car drivers or reward people who take public transport.

Car drivers can be penalised by charging them more to drive into the centre of the city. This is now done in central London, where it costs an extra £8 a day. The result has been a fall of about 20 percent in vehicles driving into the centre. This has met with approval from Londoners, most of whom did not drive into central London to work anyway. Similar road pricing is also being proposed for New York and other cities.

Raising taxes on petrol also appeals to many people as a tactic for discouraging emissions. But there is a big political problem here. In countries where most people drive to work, such measures discriminate in favour of the rich. For someone who spent £30,000 on a car, high prices at the pump make little difference. For someone driving a cheap second hand car, they do. And the experience is that you have to raise the petrol tax a long way before people stop driving to work. So raising petrol taxes amounts to taxing ordinary people and the poor. As petrol prices rise, most people will decide that the government wants rich people to drive and working people to walk a mile to catch a dirty, late bus.

Moreover, as petrol and roads are taxed, the oil companies, the car companies and the trucking companies are going to go on the attack. The great carbon corporations will launch a massive campaign against "green taxes" through the right wing press. These papers will say that they are defending the common man and woman, while the environmentalists are favouring the rich. The resulting political squeeze from the top and bottom will wreck green taxes. This is not an abstract vision. It is what happened in the UK during the fuel price revolt and national blockade by truck drivers that stopped fuel price rises in 1999.

An alternative approach is to make public transport more attractive by dropping fares and building new bus and train routes. This may infuriate the right wing, but will gain the support of most ordinary people. And the evidence is that where it is tried in the rich countries, a significant minority switch to public transport. Yet positive incentives also have weaknesses. First, they still leave the rich driving faster, more easily, in privacy and comfort. Private transport remains the ideal. That makes every form of public transport harder to encourage. Second, a majority of current drivers usually will not switch. The third weakness is that everyone misses out on all the benefits of car-free roads.

In short, banning cars and supplying really good public transport will be much more popular and make much greater cuts in emissions than taxing roads and petrol.

Switching fuels

So public transport can deliver very large cuts in CO_2 emissions, *if* half or more of all the seats are filled. This is another reason it is important for everyone to go by bus and train. Then buses and trains can come every few minutes, and still run half full or more.

On top of that, public transport makes it easier to switch fuels. The drawback to electric cars at the moment is that they accelerate slowly, cannot go very fast, and the batteries have to be charged for several hours every 200 miles or so. With urban public transport none of these drawbacks matter. Moreover, urban train and underground travel usually runs on electricity anyway.

Clean electricity makes it possible to cut CO_2 emissions to almost nothing. This only works, however, if a large majority of the electricity comes from clean non-carbon sources. Burning coal in a power station and then transporting it along cables that leak electricity consumes more energy, and emits more CO_2, than burning oil in a car engine. If only 50 percent of the electricity on the grid comes from clean sources, the cuts in emissions from electric cars will be a good deal less than 50 percent. At 80 percent clean sources, though, the switch to electricity starts to make sense.

This can then be combined with changes in design and engineering that cut energy use per mile in buses and trains, to make further savings. If new vehicle design is combined with good passenger loads, emissions per passenger on buses and trains can be reduced to less than 5 percent of the current emissions per car passenger.

This does not, however, mean cuts of more than 95 percent on current total transport emissions. For one thing, many people

Solutions That Could Work Now

already go by public transport. For another, a good cheap bus and train service would encourage more people to travel. This would increase emissions somewhat. But it is another piece of evidence that good public transport doesn't mean sacrifice—it means more chances to see the world.

Longer distances

So far I have been talking about travel in cities and suburbs. The same general logic applies to transport between cities, and eventually in rural areas.

Travelling between cities by bus is now much slower than in cars. Much time is lost because you have to go into a central bus station, wait and then hang onto your patience as the bus lumbers through the traffic. The solution is urban bus and train services that take passengers quickly to hubs outside the towns at motorway and interstate highway junctions.[52] There passengers could transfer to express buses that came every few minutes and ran along the motorways. This would get the passenger there as quickly as a car now does. Crucially, electric buses would also be possible, either with overhead lines or with the driver and passengers switching vehicles every three hours.

The other, even better, solution for long journeys is government action to encourage rail travel. Trains create about half the emissions per passenger of buses. They require more investment in building new track—again that means more jobs. Many of these lines, however, can be built on the routes of old train lines that have been closed all over the world.

Most of this can be done very quickly. Within one year Delhi converted its whole public-service fleet, including motor-rickshaws, from diesel to natural gas, with wonderful results in terms of air quality. The same can be done with electric buses in one year. This is another piece of evidence that we don't have to wait until 2050 to cut emissions.

Road freight

Railways are also the key to doing something about trucks. If cars now account for 45 percent of global CO_2 emissions from transport, trucks account for 25 percent.

Road freight is more efficient than cars because it is already a form of public transport—everyone's goods are shipped together. Fuel savings through design and engineering could cut emissions by half.[53] But a change to rail freight would be even more important. The British industrial association for rail freight claims rail is eight times less wasteful. This is probably an exaggeration, but the difference is certainly substantial. What is needed here is a massive railway building programme, combined with government regulations requiring long distance freight to go by rail. Trucks would still be needed for feeding containers to the railways and driving the containers from the station to their destination. The ubiquity of containers today, however, makes this an easy job.

Here again, a strategy for deep cuts has to combine redesigning trucks, switching to rail and electrifying the railways.

Airplanes

The third major source of transport emissions is airplanes, mostly passenger traffic. This is less important than trucking—planes account for 3 percent of current global emissions, and 6 percent of UK emissions. The danger, though, is that air travel is the fastest growing form of transport. And I will deal with current debates over air travel at a bit of length here, because they allow me to make some important general points about rationing, green taxes and individual solutions.

Planes are also destructive because they emit both CO_2 and significant amounts of rarer greenhouse gases directly into the upper atmosphere. A 1999 report from the Intergovernmental Panel on Climate Change estimated that the warming effect of air travel was roughly 2.7 times the effect the CO_2 emissions alone might suggest.[54] However, the IPCC researchers looked at the *immediate* greenhouse effects of air travel. But these were effects from gases many of which decayed quickly in the atmosphere. This is why most studies of emissions look at the *long term* effects. In the long term, over 20 years and more, airplane emissions do not have much more effect than land transport. But the short term is when we need cuts.

There are real possibilities for savings through engineering and design. Boeing claim that their new Dreamliner will use 20 percent less fuel.[55] More important, airplanes now use 70 percent less fuel per passenger than they did 40 years ago. Similar savings may not be possible in the next 40 years, but cuts of 50 percent are certainly

Solutions That Could Work Now

possible.[56] Yet none of these will compensate for the growth in air travel expected over the same period.

Most of the solutions to the problem of air travel now on offer emphasise personal consumer choices, market incentives and sacrifice by ordinary people. The problem is usually posed like this:

> If you care about the planet, stop taking planes. The central problem is that flights are getting cheaper and cheaper, and thus more people are flying. People are greedy, and want their holidays. So we will have to stop new airport runways by petitioning, demonstrating, and blockading. And we will have to tax fuel and flights to make them so expensive most people don't take planes.

On the face of it, there is a lot to be said for this argument. We do have to cut flights, and we must stop new runways. It is a scandal that fuel for land travel is heavily taxed, while airplane fuel is completely untaxed. It is an even greater scandal that air travel has been left out of the Kyoto treaty.

There is, however, a political weakness to the standard argument. It is evident in the debates over a new runway at Heathrow Airport in London. Environmentalists are quite rightly against the runway. But the New Labour government are adamant that it will go ahead. Since the 1980s successive Conservative and Labour governments have presided over a relative decline in British industry. Their strategy has been to replace this by promoting London as a global financial centre linking the US with Europe and the Middle East. New Labour and the Conservatives feel they have to keep the foreign, and particularly American, businessmen and women flying into Heathrow.

But when the then prime minister Tony Blair was challenged over the expansion of Heathrow, he did not mention business travellers. Instead he said working people valued cheap holidays abroad. (And that's true—we do.) So, Blair said, the government would never have public support for cutting air travel.

That's where the political argument stands now. On one side environmentalists ask for sacrifice. Moreover, they concentrate on banning cheap flights, largely by making them more expensive. The effect would be that only rich people could fly. So New Labour can pose as defenders of working people.

This leaves most people in confusion. In cold wet Britain we want holidays. Those two weeks in the Spanish sun are a magic compensation for the boredom and frozen rage of the rest of the year at work. Cheap flights will take a family of four from Britain to Spain for £200. The train will cost £1,000 or more. Ordinary people can't afford that. Some of us also want to take our children to visit their grandparents in Bangladesh. On the other hand, we all know flights will kill the planet. So people fly, feel guilty, and sometimes try to buy forgiveness by donating money to tree planting scams.

This is a recipe for apathy, not for mobilising people. There is always a problem when the climate movement sides with the rich and taxes ordinary people. Putting a tax on home heating, for instance, makes little difference to the affluent. But in Britain many thousands of poor old people already die every winter because they turn down the heating to save money. With further taxes and the rising cost of gas, heating bills will become an ever larger percentage of their small incomes. By the same token, taxes on petrol and roads mean that the rich will drive and the poor will use expensive, inferior public transport and dream of cars. Taxing flights will put holidays and family visits out of the reach of ordinary people.

In each of these cases, a political space opens for the corporations and politicians who don't want to do anything about climate change. They can, and will, then turn the debate against climate campaigners and recruit ordinary people, the majority, to their side.

In each case, however, there is an alternative way to cut emissions and escape the political trap. And there's a solution with air travel too. As before, the solution is not sacrificing, but doing things differently. It involves cutting the extravagance of the rich, not making things more expensive for ordinary people.

The starting point to a fair solution is to ban all air travel within each continent. But instead of asking people just to give up holidays and seeing their family, the governments can build new high speed rail lines. They can subsidise the railways for people who would otherwise take cheap flights. And the long distance trains can be run like cheap flights, with booking on the internet to ensure they run full.

This will not mean sacrifice. Train travel is easier, and you avoid the long waits at airports. Trains go from city centre to city centre. They shake the body less, and the air in the cabin does not make you sick. Anyone who has travelled with children knows trains are far better than planes, because children can move.

These can't be the new supertrains that go 150 to 400 miles an hour. They are a real climate problem, because their aerodynamics mean that the drag on the train increases with the square of the wind speed. This means that very fast trains emit as much carbon dioxide as cars or more. But slower trains, like the ordinary 100 miles per

Solutions That Could Work Now

hour ones in Britain, are all right. They could get from New York to San Francisco in 30 hours, or Paris to Delhi in 60 hours. And the seats can fold up and down into sleeping compartments.

That will take slightly longer. The fair solution is to give people two days extra holiday a year.

That still leaves people who need very long hauls, and flights across the Pacific and Atlantic. The key here, though, is to limit expensive flights, not cheap ones. One small way to do this is to eliminate the various luxury classes on flights across oceans that use four or five times the space. The rich can sit a bit cramped with the rest of us.

The larger way is simple. Many long distance flights are by business people. There are not so many of them, but a few of them fly a lot. There are two possible solutions here. One is to make it a law that everyone who takes a long flight has to stay for three weeks before they come back. The other is to ration long flights so that each person takes only one return a year, and no one can sell their ration to another. Business could rely mainly on conference phone and video calls.

Again it is the combination of measures that will make the difference—change the engineering of planes by government regulation, provide cheap air travel, extend holidays, ban flights within countries and across continents, and effectively ration flights across oceans.

This is a solution that will allow us to cut air passenger miles by 80 percent, and still allow people to travel.

What to fight for now

I have now outlined plans for making transport almost carbon dioxide free. Again, however, these solutions are very ambitious. So what can we fight for here and now, that will give people confidence to try to make these solutions a reality? I have four suggestions that might work.

The first is a series of local campaigns for car-free cities. Once we win that in one city, we would have a beacon for the world. It would be a different kind of city—more human, more livable and more beautiful. Then television and tourism would spread the word. The second car-free city would be easier to win, and after that many more.[57]

The second possibility is campaigns for free, or very cheap, public transport in cities. That would take more cars off the roads and make it easier to imagine a car-free city.

The third sort of campaign is to stop airport expansion. In many places in the world there are already local campaigns against new airports or runways. At the core of these campaigns are local people who don't want the noise, or want to save their homes. There is often an insular feel to these campaigns, of people protecting their property values. This is a strength, because it gets people moving. It is also a weakness, because local activists feel less confident of the rightness of their cause. The key here is to infuse these local campaigns with the knowledge that they are fighting for the planet. The shining example is the Climate Camp for direct action right next to Heathrow Airport in London in August 2007, that united local people with climate activists from all over the country.[58]

But limiting air travel, I have argued, also depends on giving people an alternative—railways. This brings me to my fourth campaign for the UK—taking the railways back into public ownership. The result of the privatisation of the railways in the 1990s is a bad, unreliable, dangerous and expensive service. This flows from the logic of letting private companies run a national rail network. The private companies have no interest in massive investment in new track and trains. Rather their interest lies in milking a captive population of commuters by running fewer trains, packing them tighter, and charging more for each ticket.[59]

This means that public ownership is essential to railway expansion and cheap long distance travel. It also means there are people to be mobilised—railway workers and commuters.

The political ground is also easier than it might be. For one thing, the vast majority of people agree that the old nationalised company was better. But the incompetence of the private companies means that sometimes they fail. The government has already had to take the infrastructure company, Railtrack, back under public control. In the summer of 2007 Metronet, which ran two thirds of the London Underground, went bankrupt. Under a scandalous contract, the government had to pay £1.9 billion of its debts. This makes the supporters of privatisation politically vulnerable.

Public ownership of the railways and massive investment in rail would do several things at once. They would provide an alternative to air travel. They would make rail travel a cheap alternative to cars. And they would take air freight off the roads.

Solutions That Could Work Now

CHAPTER 8

Industry

ON an abstract level, industry is perhaps the easiest sector in which to cut carbon dioxide emissions. This is because emissions are concentrated in a few industries which are globally controlled by a small number of companies. So the logistics of change are straightforward. Moreover, energy efficiency in industry has not been a major concern of most companies, so there are many immediate changes which could have a dramatic effect.

The same things that make plans easy, however, are also the things that make them abstract. It is easier to make suggestions for buildings and transport because both sectors have always been dependent on government subsidies and regulation. But industrial corporations are major political players accustomed to telling governments what to do, and not the other way round.

Cutting industrial emissions by controlling the flow of materials is another opportunity with the same difficulty. Imagine what a difference it would make if the government just told some industries to make more and others to make less. To say that blithely, in a sentence, is to reveal the problems in stark form. How likely is any government to do that? This is one reason that almost all the literature on industry and climate change concentrates on energy efficiency, and on smaller changes around the edges rather than centrally tackling the problem.

So this chapter on industry reveals in acute form a problem that runs all through this book. It's easy to demonstrate that deep cuts in CO_2 emissions are possible. But the nature of those cuts means that considerable economic and political power will be mobilised against such changes.

With all this in mind, here is what *could* be done to cut emissions from industry.

In total, global CO_2 emissions from industry in 2004 were 9.9 billion tons. This was just over a third (36 percent) of the global total of 27 billion tons.[60] Two thirds of industrial emissions came from the following seven industries:

Global CO₂ emissions by industry, 2004

Cement	1.7 billion tons
Petroleum refining	1.6 billion tons
Iron and steel	1.6 billion tons
Aluminium	0.5 billion tons
Tiles and bricks	0.4 billion tons
Pulp and paper	0.4 billion tons
Fertilizers	0.3 billion tons
Total of all 7 sectors	6.5 billion tons
Total for all industry	9.9 billion tons[61]

Three industries—cement, refineries and steel—account for more than half of all industrial emissions. Cement is a problem because the main ingredient in cement is usually limestone, a mixture of carbon, calcium and oxygen. Making cement is basically a process of taking the carbon out of the limestone and emitting it into the air as CO_2. Refineries are a problem because of the large amount of energy needed to heat crude oil and "crack" it into various useful hydrocarbons—mainly petrol, diesel, kerosene, and feedstuffs for chemicals, plastics and fertilizers. Iron and steel are a problem because of the very high level of heat needed to turn ore into useable iron and steel.

However, very considerable savings are possible from control of materials—from changing the shape of industries. Let me take each of the three big emitters in turn.

Cement production accounts for 1.7 billion tons, about a sixth of industrial emissions and 6 percent of global emissions. Half of the cement capacity in the world is now in China and three quarters of it in the developing world. This cement is used to build houses—particularly tall buildings that eat energy and are intended as homes and offices for rich people in poor countries. As I argued in Chapter 6, the answer in Asia, Europe and the United States is houses and office buildings on a human scale made from traditional materials.

Failing that, or in places where cement is necessary, there are also very large savings in emissions to be made by using industrial waste in concrete, and very large savings from using gypsum instead of limestone to make concrete. Gypsum is not currently used because it is "too expensive". Yet here too, as elsewhere, too expensive means more jobs.

Solutions That Could Work Now

Petroleum refineries account for 1.6 billion tons of CO_2 a year, another sixth of all industrial emissions and 6 percent of global emissions. Most petroleum goes on transport. The changes to transport I outlined in the last chapter would get rid of almost all that petroleum use. It would still be necessary to make some feedstuffs for chemicals, pharmaceuticals and fertilizers. But a sixth of the current global refineries should be able to meet those needs.

Iron and steel again account for 1.6 billion tons a year, another sixth of industrial emissions and 6 percent of the global total. Here it is hard to think of substitutes, because making other metals also requires great heat. Aluminium, for instance, needs even more energy than iron and steel. The real saving would have to be in newer, more efficient plants, and in switching fuels. Consider the average CO_2 emissions per ton of steel in several major steel producing countries:

Tons of CO_2 per ton of steel

Brazil	1.25
Korea	1.6
Mexico	1.6
USA	2.0
China	3.0 to 3.8
India	3.0 to 3.8[62]

There is a classic method long used by any government that wants to increase energy efficiency. Pick the most energy efficient producer and insist that within five to eight years the other companies have to meet that standard. If a government puts that into law, the companies do it. This is what governments have done, for instance, with standards in refrigerators and with gas mileage in cars.

The difficulty here is that moving to more efficient steel production confronts the problem of global competition. Steel plants require enormous investment, and are central to every national economy. No overall international body controls them. The only solution here again involves national government subsidies and job creation. That is, building new steel plants, or retrofitting old ones, in ways that match the energy efficiency of the best plants in the world.

There is one more important change that can be made to steel mills. Some now heat and process the steel with carbon fuels, and some use electric arcs. If all of them changed to electric arcs this could cut emissions drastically, but only if the great majority of electrical energy already came from wind and solar power.

Motors and electricity

This brief sketch suggests what could be done with the cement, refineries and steel. Together they represent half of all industrial emissions of CO_2 and 18 percent of total global emissions. It is possible to make very large cuts to emissions from these sectors. But this would involve measures that go far beyond energy efficiency.

There are, however, important areas where traditional energy efficiency can make a large difference. Making the electricity that is used in industry accounts for 4.8 billion tons of CO_2 emissions a year, almost half the total of industrial emissions.

To understand the possibilities in cutting electricity use, we have to come at the problem sideways by looking at what industry does with productivity. Increasing productivity means that a company produces more goods using the same number of workers. This is the holy grail of capitalist management. Increased productivity enables a company to lower prices, or to make more profits, or both. Factory engineering and machine design concentrate on this goal. So does the organisation of work, and so do managers at all levels. Most years, in most countries of the world, industrial productivity rises by at least 2 percent. This is the engine driving the expansion of the sheer amount of goods—products—in the world.

The key to increased productivity is not driving workers harder. It lies in changing the design of the machines, the layout of the job and the organisation of labour. It also lies in new investment, new plant and new machines, all of which are usually more productive than old plant. These areas are also the key to energy efficiency. But engineers and managers don't pay the same attention to energy saving that they pay to labour productivity. Whenever they face a choice between saving energy and saving labour, they opt for trying to increase productivity, almost without thinking. When they do think about energy efficiency, it's an add-on. The difference is that labour is their main cost. In almost all industries energy costs are not on the same scale.

But there is still a steady increase in energy efficiency in industry. This is in part because some measures that save labour also save fuel. It is also because energy efficiency is not that hard. But because

Solutions That Could Work Now

systematic attention is not paid, there are usually "low hanging fruit"—easy energy savings to be made.

The problem is how to make companies pay attention to energy costs. There is a school of thought among engineers that holds that most of these savings would pay for themselves if only companies did them. In many cases, this is true. The work of energy guru Amory Lovins and his collaborators is full of examples.[63]

Let's take the example of motors in industry. Electricity use in industry accounts for almost half of industrial emissions of CO_2, and 17 percent of the global total of emissions. Lovins shows that the important savings are to be made in the design of electric motors and pumps. In the European Union and the US between 63 percent and 65 percent of industrial electricity goes to power motors. The IPCC estimates a third of this could be cut, and Lovins reckons much more.[64]

The majority of energy from these motors is spent on pumps that move air, fuel and liquids around the factory. Lovins has collected a raft of ingenious design ideas for saving energy on pumps. Traditionally the pipes and ducts run from one machine, wherever it happens to be in the factory, to another, wherever that is. Lovins says start instead with the least wasteful placing for the pumps and pipes, and then move the machines to fit the pipes. Small changes in the size of the pipes also make large differences. Changing the pipes then reduces the amount of energy the motors have to produce to move the heat and liquids around.

It may be true in many cases that the cost savings alone pay for these design changes. But in most cases companies don't do them. They are set up to pay attention to other things. As before, the question is how to make energy efficiency as important as labour productivity. This will not be easy.

The obvious strategy is to make electricity for private companies more expensive by taxing it. It then becomes a large cost, and companies pay attention. The UK, for instance, has tried to do this, but the tax rates have been so low that they make little difference to business behaviour. And the reason the tax rates are low is that really expensive electricity would run into massive corporate resistance.

Simple rationing would be a more effective strategy for focusing industrial attention on energy efficiency. Again this is a case of controlling the flow of materials, as with rationing cement and closing refineries. A government could decide what amount of electricity goes to various kinds of businesses. That would achieve its stated objective in a way that cautious piecemeal taxes do not. Another solution would be to look for the technology that works to limit emissions, and then to mandate it for all businesses in that industry.

However, the main strategy here is to empower the engineers and shop floor workers. This brings me to what we can fight for now.

What we can fight for now

The two obvious immediate campaigns flow from the necessity to control the emissions from cement plants and refineries. One is the fight against tall buildings made of cement I talked about in Chapter 6. The other is the fight for public transport I suggested in Chapter 7.

My main suggestion, however, is to find a strategy to empower the engineers and shop floor workers. They work at the job every day. They have vast reserves of technical understanding. If they themselves become determined campaigners for the climate at work, they will be able to make a hundred changes in every plant. I will concentrate on the British example, though many other countries are similar.

In Britain most employers already have a paper commitment to energy saving. Many workplaces have "energy champions" in each section to look for ways to cut energy use. The management then adopt those suggestions that will save them money, and ignore those that will cost them. For instance, managers suggest cutting overtime or charging for the car park. These are effectively pay cuts for employees and money earners for the employer. But management reject suggestions to cover the roofs of their new buildings with solar panels on grounds of cost. So if they are not careful, "energy champions" can easily become the defenders of sacrifice by the workers. Union representatives, on the other hand, can find themselves defending their members' interests against the climate. This makes the union activists confused and defensive.

There is a way through this contradiction. I have spoken to several workplace union meetings in Britain about global warming. In every meeting someone who was not a union activist spoke out passionately. The Trades Union Congress has a programme for training environment representatives in every union branch, just as it trains health and safety representatives. That passionate environmentalist in the meeting could be that rep, and the local union would be stronger for their passion.

Then the union environmental reps could sit in the meetings that management holds to look for energy efficiency. They could fight as representatives of both the workers and the planet—for solar panels, against car parking charges, for a free shuttle bus and against overtime cuts. In the process, they could begin to release the creativity and intelligence of engineers and workers on the shop floor and at the office desks.

The backbone to all of this, though, is to persuade unions and workers that fight against global warming is not about sacrifice, but about jobs and a better world. This means linking energy saving with a fight to protect jobs.

The radical measures I have been proposing to stop global warming would create a lot more jobs. But they would also destroy some jobs, many of which are the lifeblood of communities. Go to Flint, Michigan, or any former mining village in Britain, and you can see what happens to people when the industry goes—mass unemployment and an exodus of young people. Drink, heroin, pills, depression, domestic violence and helpless resentment take over.

This means that climate campaigners can find themselves fighting against both some of the largest corporations in the world *and* against their unions and workers. The only solution that will work here is very radical—job protection. That doesn't mean we defend particular jobs. It means fighting for guaranteed work, at the same wages, somewhere people can commute to without breaking up their family. In some cases factories could stop making cars and start making solar panels, wind turbines, buses and trains. Truck drivers could drive buses, and airline cabin staff could work long distance trains. Offshore oil workers could build and maintain wind farms.

I know this sounds visionary. For 30 years corporations have relied on the threat of unemployment to cow unions and workers into submission. Any job guarantees would create a precedent and lead people in every industry to want them. Mainstream political parties and corporations will fight tooth and nail to prevent this.

The other reason job guarantees sound fanciful is that unions have been on the defensive for 30 years in Britain and many other countries. They have fought to defend their wages and conditions and to stop hospital closures. However, union activists who are always on the defensive cannot imagine winning a fight to guarantee jobs. And yet historically the strength of unions has flowed from offensive campaigns to win new benefits and organise in new industries. This gave union activists confidence, and a conviction that they were fighting not just for themselves, but for the whole working class. Fighting to protect jobs requires that aggressive, confident mindset. This is where fighting for the climate comes in. Everywhere the people who first built strong unions carried in their hearts a vision of a different world for working people. Stopping global warming can be part of that again. Union activists can see themselves as defending their jobs and all humanity.

Any such union climate activism will also have to fight office by office, factory by factory and building site by building site. But those fights won't work without a greater vision. Engineers and shop floor workers won't fight for more efficiency if they think the result will be lost jobs. They need to be confident, not frightened. National campaigns for changing production and protecting jobs will depend on and empower thousands of shop floor activists who nag the managers to move the motor so the pipes can work more efficiently.

Clean energy or energy efficiency

This is the last of three chapters on energy savings. There is one last point to be made, which bears on the question of what we fight for and how we fight. This is that energy efficiency matters less than clean energy.

Paradoxically, I have devoted more space to energy efficiency. This is true of almost everything you can read on climate change. The reason is that clean energy solutions are simple to explain, while energy efficiency solutions are more complex and varied. This, indeed, is one of the reasons why clean energy is more important— it simpler to explain, simpler to fight for and simpler to implement.

There is something beguiling to writers on energy and climate change about the very complexity of energy efficiency. It appeals to the inner engineer in all of us, me included. The detail is fascinating. But this very complexity makes efficiency solutions more difficult. The argument becomes expert territory. That's not the territory of the sort of mass movement we need.

Another reason is that some of the most effective ways of cutting emissions I have described above involve a switch to electricity. The most important examples are using electricity for heating buildings, for powering cars, buses and trucks, and for manufacturing steel. However, as we saw earlier with buildings, the saving only matters if the great majority of the electrical grid comes from clean energy, due to the loss of energy in delivering the electricity to the user. This problem is acute with heating, because electricity is an inefficient way of making heat.

But the problem is the same, though less dramatic, with cars and buses. Burning petrol in the engine is more efficient than producing electricity at a distance.

What these examples mean is that the real savings in emissions from changing to electricity only kick in when 90 percent or more of the grid comes from clean energy. That's another reason why clean energy is so important.

Moreover, clean energy can solve the problem *quickly*. Given the political will, wind turbines, solar power and the rest could cover the world in five years. Clean energy is also a *simple* solution. It's easy to understand, and therefore easy to fight for. No one gets lost in technical arguments, and regulation is easy.

So of course we have to fight for both clean energy and energy saving. But clean energy matters more.

CHAPTER 9

Solutions that Won't Work Now

THIS chapter is about the solutions which won't work. There are several reasons to be clear about these.

First, it is possible to write as if the alternatives for clean energy are so cheap and abundant that there is no problem. The difficulty with this approach is it's not true. Some proposed solutions don't work yet, some will have little impact, some will have devastating consequences and some will make it more difficult to halt climate change. If you mislead people, you will be caught out. That will demobilise people and make it harder to fight for climate action.

Second, very large sums of money will have to spent on clean energy and energy saving. If this money goes on expensive alternatives that won't solve the problem, it won't be available for solutions that can solve it.

Third, almost all the questionable solutions are backed by corporate interests in order to prevent changes that threaten them. For example, the coal and industrial corporations back carbon capture and storage so they don't have to cut their emissions.

The final reason for looking at the solutions that don't work has to do with the political implications of the necessary changes. If none of the proposed solutions to climate change worked, there would be no hope. If all of them worked, we would not need drastic change in the shape of the world economy. The actual situation is that some solutions will work now, and some won't. The ones that will work, like public transport and concentrated solar power, will challenge corporate power and require large changes in the shape of the world economy. The argument running right through this book is that we have to change the world to save the Earth. So the whole argument of this book depends on the technical detail in this chapter.

One final point—I hope I'm wrong. It is possible that technological progress will prove that biofuels, hydrogen or carbon capture and storage will go a long way to halting climate change. This will be wonderful if it happens. But for the moment, it doesn't look as though they will work.

Biofuels

I will start with biofuels and then hydrogen. In an earlier chapter I outlined ways to replace cars, trucks and planes with buses and trains. On one level, these changes are not a technical problem. This does not mean they will be easy to win. The biggest political problem isn't the greed of car drivers, but the power of automobile corporations, airlines and oil companies.

No one should underestimate the power of these companies. Car companies in particular can see the writing on the wall. They are now promoting two solutions—biofuels and hydrogen. Both these solutions protect the dominance of cars, prevent a move to public transport, and won't work now to stop global warming. Biofuels also provide an alibi for airlines.

"Biofuel" is a term used to describe any plant, tree or animal product that can be burned for energy. Asian villagers traditionally use wood and cow dung for cooking—these are biofuels. So are the willow trees or sugar cane waste used in power stations. The most important biofuel now is ethanol (alcohol) made from corn (maize), sugar cane and soya beans. This ethanol can be mixed with petrol, or can power cars on its own.

The standard argument for biofuels is this:

Biofuels are full of carbon. When they burn, they release that carbon into the air. But biofuels differ from coal, oil and gas because they are replaceable. The farmer is planting more corn even as the motorist is driving the car. That new corn takes the same amount of CO_2 out of the air as the car emits burning ethanol. Then the corn makes ethanol again, the CO_2 goes into the air, and returns to another generation of plants. There is no net addition of CO_2 to the air. And the oil that would otherwise be burned stays in the ground.

This persuasive logic does in fact justify much of the traditional village use of biofuels for heating and cooking. But it does not work for ethanol fuel for cars. Ethanol has two drawbacks. One is that producing "carbon neutral" ethanol in fact releases more CO_2 than burning oil. The Minnesota University ecologist David Tillman and economist Jason Hill put it this way:

Because of how corn ethanol is currently made, only about 20 percent of each gallon is "new" energy. That is because it takes a lot of "old" fossil energy to make it: diesel to run tractors, natural gas to make fertilizer and, of course, fuel to run the refineries that convert corn to ethanol.

If every one of the 70 million acres on which corn was grown in 2006 was used for ethanol, the amount produced would displace only 12 percent of the US gasoline market. Moreover, the "new" (non-fossil) energy gained would be very small—just 2.4 percent of the market. Car tune-ups and proper tire air pressure would save more energy.[65]

That means ethanol in American cars would make people think they were doing something, while actually doing nothing.

In much of the rest of the world, biofuels damage the climate. Let's take ethanol from sugar cane, which now powers much of Brazilian transport. If this sugar cane is grown on land that was already cleared and has traditionally been used for crops, there is a reduction in CO_2 emissions compared to petroleum diesel. The saving is only 40 percent, because carbon fuels are still needed to make the fertilizer and refine the cane. But Brazilian savannah and forests are being cleared for cane, releasing large amounts of CO_2 from plants and soil. The result is that farming savannah for ethanol releases more CO_2 than burning petroleum diesel. Cutting down forests creates a drain from the soil that lasts and lasts, and means that the sugar cane ethanol will put 50 percent more CO_2 into the air over the next 20 years than petroleum diesel would.[66] With the rising price of biofuels, more and more Brazilian rainforests will now also be cleared to make ethanol from soya.

The situation with palm oil, widely used to make diesel, is even worse. Indonesia now produces 43 percent of the world's palm oil, mostly on land carved out of the rainforest. This dries out the peat. A recent report from Wetlands International estimates that the dried peat bogs emit 0.6 billion tons of CO_2 a year. The dry bogs are also very vulnerable to fires, which emit roughly 1.4 billion tons of CO_2 a year. If these figures are right, growing palms for biofuels in Indonesia alone accounts for 2.0 billion tons of CO_2 emissions a year—more than the global total for air conditioning, or trucks, or cement or steel plants.[67]

These figures are estimates and may be exaggerated. Even so, it is still clear from Hooijer's study that the CO_2 emissions as result of growing the palms, refining the oil and shipping it across the world

Solutions That Could Work Now

mean that palm oil diesel in Europe has caused ten times the CO_2 emissions of the equivalent amount of petroleum diesel.[68]

So biofuels don't save emissions. They will make the effects of climate change worse, because of a basic problem with all biofuels. Given a choice between using land to grow food and to grow automobile fuel, the market will push farmers to make the fuel every time. That will reduce the amount of land for growing crops, push up the price of food and leave some people to go hungry. Biofuels and failing rains together will create famines in order to feed gas to cars.

Biofuels are already crowding out food. The US government under George Bush has swung hard to encourage production of ethanol from corn. Bush justifies this as a way of fighting climate change. The car companies support this. So do the oil companies, because the ethanol is mixed with petrol. The petrol is still mainly oil, but hey presto, it is somehow now green. Agribusiness, which controls most US farming, also sees a gold mine in biofuels for cars.

By 2006 there were over 100 plants for processing ethanol from corn in the US, and many more under construction. The US is both the world's largest producer and exporter of ethanol. At the end of 2006 the UN's Food and Agriculture Organisation reported that price of corn had doubled, both in the US and globally. The FAO said this was driven by the switch to corn in the US. As people switched to other grains, the price of rice and wheat was also rising. So was the price of chicken and pork in the US, as both are mainly fed on corn.[69]

Mexico was hit hardest. Tortillas from corn are the staple diet there, and the cost had doubled. In February 2007, 75,000 people demonstrated in Mexico City for cheaper tortillas in a protest organised by the unions and the opposition party.[70] But this is a global problem, and it will get much worse. The American, British, European Union, Brazilian and Indian governments are all encouraging biofuels.

What will make this worse is that global yields for food crops are expected to fall with the changing weather patterns that accompany global warming. Drought will be widespread, the South Asian monsoon is likely to be disturbed, and the glaciers of the Himalayas are already melting. It is true that some areas will get more rain. But even there much of the rain will fall out of season, or in intense bursts that create run-offs and floods instead of sinking into the soil. If this is combined with a squeeze on food supplies from biofuels, we will have famine to feed cars.

So biofuels are not a solution. We will have to replace cars and planes with buses and trains.

Hydrogen

Biofuels are pernicious. By contrast, the only problem with hydrogen is that it doesn't work yet.[71]

Many people are drawn to hydrogen because it seems to offer a way to continue using cars without causing global warming. In practice, hydrogen has been used as a blind by the car industry. The car industry in the US is constantly announcing that it is making hydrogen cars, without ever actually offering them for sale.

Many people are also impressed because hydrogen fuel cells can run a car and emit no carbon dioxide at all. Instead they just mix hydrogen and oxygen to make water vapour, heat and energy. You can't get cleaner than that.

This is misleading. Hydrogen is not a fuel. It's a way of storing energy, like a battery. Hydrogen is almost never found in pure form. To make hydrogen (H) for a fuel cell, you first have to use electricity to separate the H from water (H_2O). The hydrogen then stores the energy from the electricity. When the hydrogen burns it releases the energy and combines with oxygen to make H_2O.

At the moment almost all hydrogen is separated from water by using electricity from coal and gas. So there is no reduction of CO_2 emissions. In fact, emissions increase because hydrogen must be compressed before it is transported. That compression requires a great deal of energy, which again comes from coal and gas.

Even when hydrogen is made with clean electricity there are still serious problems. It has been suggested that hydrogen could be used to transport the energy from concentrated solar power. An impressive 2006 study by the German Aerospace Agency, however, found it would take three quarters of the solar energy produced just to compress the hydrogen for transport.[72]

Moreover, there has been little technological progress with hydrogen. So unlike wind and solar power, hydrogen has been around for a long time without getting much cheaper. In 2003 hydrogen made with carbon fuels cost 30 times as much as the equivalent amount of petrol. Hydrogen from renewables was even more expensive. Joseph Romm, in his trenchant book on hydrogen, points out that "to replace all the gasoline sold in the United States today with hydrogen from electrolysis would require more electricity than is sold in the United States today".[73]

There are also safety problems. Hydrogen leaks easily, and is flammable at much lower temperatures than petrol. The static electricity from turning on a mobile phone, or sliding your bottom over

Solutions That Could Work Now

a car seat, can start a hydrogen fire. On the plus side, the fires are not that hot.

And yet, and yet. Hydrogen is almost the only way engineers have found to store energy. And it makes more sense installed in homes, because hydrogen fuel cells produce both heat and power. So hydrogen does not work yet, and will probably take a long time to develop. If it does work one day, that would be wonderful. But we need to cut carbon emissions long before that day arrives. At the moment hydrogen cars are a distraction from solutions that work. We still have to replace cars, planes and trucks with buses and trains.

Carbon capture and storage

"Carbon capture and storage" (CCS) is also called "carbon sequestration". The hope is that it will allow power plants and factories to keep emitting CO_2. But the carbon will be captured as it goes into the air and stored underground. This will allow industry and coal and gas companies to continue working more or less as normal.

Carbon capture works like this. The carbon dioxide exhaust leaves the power plant or factory through a pipe or tower. As it does so the exhaust passes through a filter, called a "scrubber", which traps the CO_2. That CO_2 is stored temporarily in barrels that are then poured into underground caverns for permanent storage.

Unfortunately, it isn't working and it won't work.[74] The International Energy Agency estimates that for CCS to have any effect on the climate there will have to be 6,000 projects each putting a million tons of CO_2 a year into the ground. Right now there are three big projects. None of them are attached to a working power station. Nowhere in the world is there a power station of normal size scrubbing and saving carbon.

One reason is the cost. Fitting the technology would double the cost of building a power station. The technology sucks energy—estimates are that 10 to 40 percent more coal would have to be burned just to power the scrubbers. Then there is the problem of transporting the CO_2. For that it has be pressurised, at great cost in energy. Transport will probably be by pipelines, which have not been built, and are likely to be very expensive at distances over 100 kilometres. But most power stations are being built at considerable distances from places where CO_2 might be injected into the ground. That's where the problems really begin.

We have no experience of storing large quantities of CO_2 in underground. Indeed, we have no experience of storing large quantities of *anything* underground. There may well not be anything like enough space available underground. No one knows how long the gas will be held underground, or what the rate of leaks will be. Annual leaks of 1 percent will return most of the CO_2 to the air in less than a century. Indeed, the power industry globally is refusing to develop CCS unless governments agree to guarantee their insurance against liability. In the United States the industry has successfully insisted that the federal government pay all the insurance bills and pass laws making it impossible to sue the power companies for leaks or other consequences of CCS. No business believes the technology can be trusted.

Moreover, almost no one expects the technology to be operational before 2030—after we need to have stopped climate change. Many expect it to begin working after 2050. So it is an alternative that does not work, instead of one that does. And CCS is taking money away from cheaper, reliable energy. In 2008 the US government budgeted three times as much spending on CCS programmes as on all other renewable and energy efficiency programmes combined. For 2009 they propose to spend four times as much—$624 million as opposed to $146 million.[75]

It's worse than that, though. CCS is being sold as "clean coal". What this means is that new power stations are said to be "CCS ready". That means they can be retrofitted with scrubbers, pipelines and storage caverns later. But any power station, in theory, is "ready", because any can be retrofitted at some cost. The cost for almost all the new power stations will be prohibitively high. The effect is that new coal fired power stations are being built all over the world under claims that CCS will work, but CCS will not in fact be fitted to them. Where it is fitted, it will be long after it is too late.

Carbon capture and storage does not work, won't work in time, and is a cover for increasing coal fired power stations, the most destructive way of producing electricity.

Peak Oil and Gas Cliffs

There is one more suggested solution that is definitely not being promoted by the car and oil companies, but still won't work. Many people now see hope in the world's dwindling supplies of oil and gas. The problem here is that coal, which is worse, will replace oil and gas.

Solutions That Could Work Now

"Peak Oil" is real enough.[76] The idea was first suggested in 1956 by M King Hubbert, an American petroleum geologist working for Shell. Hubbert described how the oil reserves in any country went through a cycle. In the early years discoveries came quickly, and the annual production of oil increased. Then, when most of the big fields had been found, production began to level out. Finally the oil would become increasingly difficult and expensive to extract. The high point of this cycle is now called "Hubbert's Peak".

Hubbert turned out to be right. It is difficult to remember now, but the United States was once the world's largest oil producer. It passed Hubbert's Peak in 1970.

The oil companies have been hostile to the idea of Peak Oil, so it is difficult to judge when we will pass Hubbert's Peak globally. Honest figures on oil reserves are a closely guarded secret. The major oil companies all give numbers for the reserves they control. But their share prices depend on these estimates. In 2004 the senior management at Shell were caught exaggerating their reserves by more than 20 percent, and the share price fell dramatically. It is widely believed in the industry that many other companies also exaggerate their reserves.[77]

The major oil producing countries also provide unreliable numbers. Countries in the Organisation of the Petroleum Exporting Countries (OPEC) are allowed quotas for the amount they can produce each year based on their estimates of reserves. When this rule came into operation, most of the major producing countries promptly raised their estimates substantially. Even more telling, those estimates have now remained the same for 20 years, which is not possible.

Controversy over Peak Oil continued for many years. The supporters of the theory were mostly a minority of rogue petroleum geologists, and those who rubbished it were loyal to the oil companies and the governments of the petroleum states. In 2005 the question was decisively settled by Matthew Simmons's brilliant and influential book *Twilight in the Desert*. Simmons is the head of an investment bank in Houston specialising in the oil industry. He also raised money for George W Bush's presidential campaigns. Elite circles did not see Simmons as having an axe to grind.

At least as important, though, were the conceptual rigour and beauty of his study. There is common agreement that Saudi Arabia has by far the largest oil reserves on the planet. Simmons looked at the texts of hundreds of unpublished papers by engineers working in Saudi Arabia given to conferences of the Society of Petroleum

Engineers in the years from 1961 to 2004. These papers were honest, and designed to show other engineers the professional skills of the authors. Simmons shows that over 43 years those papers overwhelmingly dealt with the sort of technical problems a petroleum engineer faces when they are trying to nurse production from a declining field.

However, we can only guess the point at which the world supply passes its peak. Informed guesses range from 2006 to 2015. The sustained oil price rises of 2007 and into early 2008, however, are suggestive.

The hope this provides for halting climate change is simple: *As oil stocks decline, the price will rise. Companies and individuals will switch to wind, solar or hydrogen power simply to save money.*

This would be wonderful, if it were so. Unfortunately, there is enough coal to last the world for 200 to 300 years. Coal can easily be turned into a gas that works like petrol. That fuel will be more expensive, which may deter some users. But gas from coal will create much larger emissions than the equivalent amount of oil. This is partly because of the energy cost of making the gas, but mostly because coal emits much more CO_2 than oil. And the less oil is available, the more people will use coal to heat their houses.

Moreover, rising oil prices are already leading to increased investment in the oil heavy "tar sands" of Canada and Venezuela. It requires considerable energy, and emissions, to extract the oil from the tar sands, and that oil itself is high in carbon emissions.

A process similar to Peak Oil is also happening with natural gas. Here the geology is different. In an old, half drained field the oil becomes progressively more difficult and expensive to extract. In a natural gas field the gas comes out steadily and then stops abruptly and almost completely. So while oil supplies tail off in a curve over many years, gas supplies hit a cliff.

The UK is now hitting this cliff with North Sea gas. At some point in the near future the same will happen in North America. The other gas fields across the planet have much larger reserves, and it it will be 30 years or more before the Earth begins to run out.

Meanwhile, the consequences of any gas cliff are even worse than with falling oil supplies. There is increasing interest in the US and Japan in Liquefied Natural Gas (LNG). But LNG burns a lot of carbon to liquefy the gas under pressure and then ship it.

More ominously, the main use for natural gas presently is in power stations, and coal is the main competitor there. Natural gas is the "cleanest" of the three carbon fuels, and coal is the "dirtiest".

Solutions That Could Work Now

CO_2 emissions in the UK fell dramatically in the 1990s because Margaret Thatcher had destroyed the local coal industry, and the power stations switched to gas. With the exhaustion of North Sea gas, they are going back to coal, and UK emissions are rising.

So Peak Oil and Gas Cliffs are not good news. In the next generation peak oil will provide massive profits for the oil companies. Competition for the existing oil will intensify, and so will the pressure for war. And peak oil and gas will increase CO_2 emissions, not reduce them. In short, peak oil and the market together won't solve the problem. We still have to replace coal with wind and solar power.

Dams

Hydroelectricity from dams is another solution often proposed. Dams hold back the water, and then allow it to fall from a height through a tunnel. The falling water turns a dynamo and makes electricity. Dams currently generate a great deal of electricity in many parts of the world, and seem to do so without CO_2 emissions.

There are several problems, however. The most important is the emerging evidence that vegetation decays underwater in the lakes behind dams and creates large amounts of methane.[78] And remember, methane is a much more potent greenhouse gas than CO_2.

Dams also have environmental and social consequences rather like the effects of climate change. In Egypt, for instance, the Aswan Dam has blocked the silt that used to replenish the soil each year in the Nile Delta. This has happened with many other dams.[79]

Moreover, most new big dams will be in poor countries. In all cases these dams displace very large numbers of farmers and townspeople. These people always receive little or no compensation. This is why dams seem cheap—farmers pay for them with their family lands. Then they become, in effect, flood refugees. One of the main reasons I oppose climate change is that it will create many flood refugees. A recent report by Christian Aid estimates that there are 155 million refugees and displaced people in the world. Of these, 25 million have fled war and conflict, 25 million have fled natural disasters and 105 million have been displaced by development projects, mostly dams.[80] And the largest social movement in India in the last 20 years has been the resistance by millions of poor farmers and labourers to the building of a string of dams in the Narmada valley.

On the other hand, small "micro hydro" projects without big dams have few drawbacks. Many people are drawn to these projects because small is beautiful. But small is also small. Micro hydro won't make much difference globally.

Nuclear power

The last major solution to global warming often proposed is nuclear power. Nuclear power is the heroin of alternative fuels—very tempting and very bad for you.[81]

There are longstanding objections to nuclear energy, and they remain valid. But here I am going to concentrate on the major objection—nuclear power will make it harder to stop climate change.

The temptation of nuclear energy is easy to see. The US emits 20.2 tons of CO_2 per person annually, the UK 9.6 tons and France 6.7 tons. The difference between the UK and France is almost entirely that France supplies 80 percent of its electricity from nuclear power plants.

Globally, most of the current power stations are coming to the end of their safe lives. Governments all over the world are now planning a new generation of nuclear power stations. The governments that already have nuclear weapons want new power stations for the same reason they were built in the first place. This is because civilian nuclear power makes nuclear look clean. Without power stations there would only be bomb factories. And no one wants them in the backyard. If governments close down nuclear power stations, that will encourage the movement to ban nuclear weapons.

Many other countries without nuclear weapons are also considering building nuclear weapons after the Iraq War. Their leaders noticed that the US invaded Iraq and did not invade North Korea. The difference, they suspect, is that North Korea had the bomb, so they would be well advised to get one. That means building nuclear power plants. All the countries that have built nuclear bombs since 1950—Israel, India, China, Pakistan and North Korea—have started with civilian nuclear power stations.

All the governments planning new nuclear power stations now say they are doing so to prevent climate change. Some environmentalists have gone along with them. Yes, they say, nuclear energy is dirty and dangerous, but we have to risk it because of climate change. Some of these environmentalists are often quite right wing to begin with, and they do not represent the green mainstream.

Solutions That Could Work Now

Greenpeace, Friends of the Earth and the other major green organisations remain adamantly opposed to nuclear power.

The crucial reason is that nuclear power will make it harder to stop climate change. For one thing, there is nowhere near enough uranium in the world to replace carbon fuels. The proposed new nuclear power stations will only make a small dent in CO_2 emissions. Second, even without the long wait for planning permission, nuclear power stations take more than ten years to build, while solar and wind power could carpet the world in five years.

Third, nuclear power stations are very expensive, and can only be built with large government subsidies. Governments are now under pressure to build renewable energy—wind and solar power. The governments will say they are building nuclear plants to stop global warming, and that makes them "renewable energy". The British government is already doing this. The government budget that was spent on wind and solar power will then be diverted to nuclear subsidies. Any laws that insist electricity companies have to take a certain percentage of renewable energy will include nuclear power as a renewable. That will also mean less wind and solar are built and nuclear will crowd out wind and solar power. But wind and solar power are cheaper and faster to build, so nuclear power means more emissions than the alternatives.

Fourth, at some point there will be a serious nuclear accident somewhere in the world and the nuclear power stations will be closed. Then the money that should have been spent on safe renewables in the first place will have been thrown away.

I am not the only person who thinks there will be such an accident. The head of every major business in the world agrees with me. No insurance company will write a policy covering a nuclear power station in the event of a disaster. No investors anywhere will build a nuclear power plant unless the government has passed a law saying that they will not be liable for the results of a serious accident. No banker will loan the money. No power company will build a reactor anywhere in the world without such a law. This is why the US and many other countries have such laws.

One reason for thinking such an accident likely is that they have happened before. There was almost a serious meltdown at Three Mile Island in Pennsylvania in 1979. In 1986 there was a serious accident at Chernobyl in the Ukraine, then part of the Soviet Union. The Soviet government sent in 600,000 people to clean up Chernobyl. Most were conscripts, and all were exposed to serious doses of radiation. A recent report by Greenpeace and the health ministries of Ukraine and

Belarus estimates that at least 30,000 people have died from cancers due to the accident, and possibly as many as 500,000.[82]

The result of Three Mile Island and Chernobyl was that no new nuclear power stations have been built in the US since, and very few in Europe. The nuclear power plant managers assure us that with new technology it will not happen again. You can believe that if you want to. But the accident at Chernobyl happened because the technicians were too afraid of their bosses to report the problem immediately. It is hard to imagine a technology that spans thousands of very large, very complex plants around the world operating 24 hours a day that does not lead to such human error somewhere. Moreover, the urge will be strong to cut corners in an industry now run for profit.

Of course there are also reasons for opposing nuclear power that have nothing to do with climate. Nuclear energy is poisonous at every stage of the process—in mining, transportation, manufacture and in storing the waste. Nuclear power is also an alibi for nuclear weapons. I am against climate change because I want to prevent massive human death. Nuclear war would not only mean that—it could mean the end of life itself.

But the key to my argument here is that nuclear energy will make it harder to stop climate change. Spending on a small number of nuclear plants will crowd out spending on wind and solar power that could generate enough power to stop climate change. All that spending will be wasted after an accident.

Yet nuclear power does provide an example of how, when governments really want to do something, they are willing to spend very large sums of money without any regard to the market. It is not possible to estimate the total global subsidies to nuclear weapons and nuclear power over the last 60 years. But it was more than we need to spend to stop climate change.

The difference is whether your priority is life or death.

The alternatives

This chapter has looked at a series of alternative ways to limit emissions. Many of these solutions are popular with business and governments because they hold out hopes of stopping climate change without having to change the way things are. But all of them either don't work now, will have little impact or will make climate change worse.

Solutions That Could Work Now

The argument about these alternatives is technical. Carbon storage, for instance, stands or falls on its merits. But the implications of these arguments are not just technical. Carbon storage, hydrogen cells, nuclear power and biofuels are not the solutions. This is very unfortunate. These are the solutions which would disturb established powers least, and would therefore be easiest to push through. But reality is what it is. If those solutions are not going to work in time, we have to go for the solutions which will work.

We can rewire the world. We have the technology, the money, the workers and the brains. We can do it fast, we can do it every country on earth, and we can end poverty in the process. But the solutions that will work will also require us to challenge the power of the corporations. Part Three will explain why.

CHAPTER 10

Methane and Forests

THIS is the last chapter on solutions that could work now. The previous chapters have been about reducing CO_2 emissions by replacing coal, oil and gas. This one is about methane and forests. As with previous chapters, the lesson is that the cuts that matter will come from government regulation. I also suggest two more demands that will be particularly important for climate activists in Asia and Latin America.

Methane

Methane always comes as a footnote in discussions of how to stop global warming. It's more important than that, though. As we saw in Chapter 1, this is because methane emissions can be cut rapidly and any cuts have a powerful immediate effect.

To recap, during the short period before it breaks down, methane has a much stronger warming effect than CO_2. Over the average 12 years it takes to break down, methane is about 20 times as powerful as carbon dioxide is over 100 years or more. It is also useful to think of the difference in another way. If we could cut CO_2 emissions by 80 percent next year, that would only stabilise the amount of CO_2 in the atmosphere, because it would take so long for the accumulated CO_2 to disappear. But if we cut methane emissions by 80 percent next year, that would cut the total of methane in the atmosphere by 80 percent within 12 years. This is because the accumulated old methane would be gone.[83] In short, cutting methane emissions is less important than cutting CO_2, but can have a rapid effect.

Methane emissions are already falling slightly, while CO_2 is still rising. This is because methane emissions are easier to cut than CO_2, and the necessary cuts challenge fewer corporate powers.

In discussing CO_2, I was able to describe the share of emissions from different sources. It's impossible to list the same sort of per- centages with methane. CO_2 emissions come from burning coal, oil and gas, and their sales are counted anyway. Methane emissions are all leaks and difficult to count. So the highest estimates of annual methane emissions are five times the lowest estimates. Such a large discrepancy actually means that no one really knows. In fact, the only reason we know methane emissions are declining is that we can measure the total level of methane in the atmosphere, and that's falling.

The best guess is that landfills and natural gas leaks are the largest sources of manmade methane emissions. In the past the natural gas found in many oil fields was simply allowed to escape or burned off. With the rising price of gas, this is now happening much less, and is technically easy to eliminate entirely.

Gas leaks are a more stubborn problem, because repairing leaks costs money. One estimate is that half of all methane emissions come from gas leaks. The leaks happen especially in pipelines from the gas fields and in pipes that carry the gas to houses. One solution is to line the pipelines with sealants that close the leaks. In some cases, this sealant can be injected down the pipeline. In other cases the pipeline can be opened and insulated, or insulated on the outside. Domestic gas pipes, however, have to be dug up and insulated or replaced. Again this means jobs digging up roads.

In short, it's technically easy to repair gas leaks, just expensive.

Using renewable energy would eliminate almost all methane leaks. The gas would simply stay in the ground and not go down the pipe.

The other major source of methane emissions is from landfills. The solution is technically easy. Pipes channel the methane to the surface, where it is burnt off to make heat and sometimes electricity. This is now commonly done and makes the landfill owner a tidy profit. This is probably why methane emissions are falling.

In India, China and other poor countries landfills are often small and scattered, and there has been less methane burning. But there is no reason why the technology can't be widely used in Asia, with the same sort of profits.

What all this means is that the majority of methane emissions can be eliminated in a few years. That will have an immediate effect on global warming. Given the political will, all this takes is government regulation of landfills and gas pipes.

Emissions from rice fields and beef cattle are less important, but harder to cut. Changes in the diet of cattle can reduce the amount of time they spend digesting. New strains of rice, careful drainage of fields and changing the timing for transplanting seedlings can all cut emissions.

Forests

Methane is important, easy, and receives little attention. Forests receive a lot of attention, most of it misplaced. Protected old tropical rainforests matter a lot. "Sustainable forestry", on the other hand, does little good and often a great deal of harm.

Human action has put large amounts of CO_2 into the air by cutting down forests. As with methane, these emissions are hard to count, and estimates vary widely. The impact of cutting down trees may be almost a third of the total from burning carbon fuels, and it may be less than a tenth.

The old forests of the earth, especially the tall dense tropical rainforests, hold large amounts of carbon. Ranchers and farmers are cutting down some of these forests, but timber and biofuel companies are cutting down more. Rainforest death releases not only carbon from the trees, but also carbon and methane stored in the soil. The rainforests are an absolutely necessary defence against global warming.

The pine forests of the north hold much less carbon. Recent research also suggests that the dark northern forests may even speed up global warming because they absorb more light than snow and ice on bare ground.[84]

In the long term, it's much more important to stop burning coal, oil and gas than it is to protect forests. There is far more carbon in the earth's reserves of coal, oil and gas than there is in the trees and soil of all the forests in the world. But defending the forests can make an important difference in the short term. As we saw earlier, Indonesia's emissions from deforestation alone are equivalent to a third of the total emissions each year from every form of transport on the planet.[85]

Defending rainforests matters. There is no market solution—just leave them alone.

The market solution for trees is called "sustainable forestry". It's a con. The logic goes like this:

Wood is a traditional biofuel. You burn the whole of the tree to make energy. That puts carbon dioxide into the air, sure. But at the same time the forester or logging company is planting new trees. As they grow, they take the same amount of carbon out of the air. The carbon is recycled. Burning wood does not put any more carbon into the air. In fact, as waste land is converted to forest, more and more carbon is stored.

Solutions That Could Work Now

Only what happens does not work like that logic. Sustainable forests are often grown where old forests were. The old forests, especially in the tropics, were tall and dense with biomass and undergrowth. They are cut down, and in their place sustainable foresters plant trees—usually pine, poplar, aspen, willow or eucalyptus—which grow to harvestable height very quickly. Within ten years those trees are cut down. Moreover, these are monoculture forests where one kind of tree is grown for miles. They are deserts for wildlife, and planted in ways that minimize undergrowth, so they are easier to cut. Eucalyptus also drops sap that kills all plants beneath it.

In short, sustainable forests are much less dense than the old forests, so they hold less carbon. Moreover, they are cut down quickly. If they were left to grow and grow, then the carbon would be locked in the trees for decades. Of course the trees would finally fall or burn in 30 to 50 years. But the crucial period for halting global warming is the next ten to 30 years.

Here again it is helpful to understand the forces behind the idea of "sustainable forestry." The first is the logging companies. Some are owned by rich men in Thailand and Indonesia. But most logging is done by massive multinationals. Sustainable forestry acts as a cover to justify their destruction of the rainforests. These corporations are not powerful in global terms. But they are locally powerful in Thailand, Indonesia, the Brazilian Amazon, the north western US and the other forested regions.

Sustainable forestry has also entered the discourse of climate activists in two ways. One is that during the negotiation of the Kyoto agreement Norway, Canada and other northern countries argued successfully that they should not be asked to cut their CO_2 emissions by as much as other countries because they were protecting their northern pine forests. The other is the large number of "carbon offset" companies that take money from guilty Western air travellers for what are often imaginary tree planting schemes.

In both cases, it looks superficially green to try to solve a problem by growing green trees and then "recycling" them. In fact it is not. The solution here is to be deeply green—don't cut down the forests, and plant new ones and let them turn to jungle.

What can we fight for now?

The answers here flow clearly from what I have already said about the problems.

For methane emissions, the most important campaigns are actually the ones for renewable energy. The less gas we use, the less leaks.

The other campaigns are local campaigns to persuade municipalities to insist on burning off the methane from landfills. These campaigns will be particularly important for activists in Asia.

With forests, the central campaign is to protect the rainforests that remain. This is a key campaign for climate activists in some major countries—Indonesia, Malaysia, Thailand and Brazil. To succeed, the campaigns have to be built in the forested countries, by the people around the forests. Jobs would be lost in timber mills and palm oil plantations. But livelihoods would also be created in the forest, in hunting and gathering, and in tourism. Some of these could be livelihoods for the small embattled bands of indigenous people that dot the world. Most would be livelihoods for those like the Brazilian *caboclos*, people from every background in the world who live in and on the edges of the forest. The point is not to develop national parks. In the developing world these are typically expensive playgrounds for the global rich. The local people camp at the edges. They are hungry for the land to feed their families, and they are shot dead as poachers when they hunt in the forest. Instead of national parks, these would be forests for the local people to live in and use for themselves.

In the past many rainforest campaigns have foundered on the different agendas of international NGOs and local activists. The campaigns in the Brazilian Amazon in the 1980s are a good example. There the local *caboclos* wanted to be able to continue to make a living in the forest, and regarded the ranchers and logging companies as the enemy. The international NGOs wanted to preserve pristine nature, and regarded international lobbying of Western corporations and governments as the main tactic. The NGOs had the money to pay local campaigners, and thus determine the political direction of the local campaign. In the process, the paid activists lost the support of local working people, and the campaigns achieved little.[86]

Any successful struggle to stop global warming is going to be built on the idea not of preserving the world *from* humanity, but *for* humanity.

Solutions That Could Work Now

PART THREE

WHY THE RICH AND POWERFUL WON'T ACT

CHAPTER 11

Neoliberalism and Profits

PART Two was about how we can stop global warming. But to lay out the solutions is to lay out the problem. It's not the technology. It's politics.

There are three central political problems. The first is that the necessary changes mean massive public works and government regulation. This flies in the face of government and corporate policy globally over the last 30 years. That's what this chapter is about.

The second problem is that doing something about climate change is a direct challenge to the power of the oil, coal, gas and car corporations. That's the next chapter. Finally, Chapter 13 explains how international competition sets countries against each other in the world economy. Yet governments have to somehow come together to regulate their economies together.

There is a pessimistic conclusion that can be drawn from these political problems—it will not be easy to win the changes we need. That's true, but if we are to achieve anything we need to face the truth squarely.

There is also a positive conclusion to be drawn. I show why corporate leaders and mainstream politicians will have great difficulty acting. It follows from this that that climate campaigners need to look elsewhere for the power that can save the planet. That means looking to the majority of ordinary people, and to the movements for social justice. Indeed, social justice activists are fighting against the same corporate power and public neglect that make it so difficult to stop climate change. What I call "climate justice"—an alliance between climate activism and social justice—is not just a politically correct sentiment. It is a necessity if we are to halt global warming.

Neoliberalism

To halt global warming we will need massive government action. But over the last 30 years the corporations and governments have transformed the world economy. They have privatised, cut public services and taxes on the rich, and torn up environmental and business regulations. This transformation is called "neoliberalism". The word comes from the old 19th century meaning of liberal—someone who thought government should stay out of the way of business.

The governments and corporations have done everything they can to persuade people there is no alternative to the global market, and no way to fight back. The governments and corporations don't want people to give up those ideas. Yet these ideas would be shattered if governments acted to save the climate, because everyone would then know there is an alternative to the market.

This has produced a strange contradiction I will explore in later chapters. George W Bush and the oil and car industry executives who want to do nothing are in a minority among the rich and powerful. Most of the capitalists and politicians know they have to act. They talk constantly about action. They signed the Kyoto agreement, and want new agreements. But the agreements they sign are riddled with market loopholes and scams. And they keep expanding roads, airports, steel, oil and cement. They can't take the big measures that need to be done.

Yet the rich and powerful are not simply lying about wanting to do anything. They are themselves caught by neoliberalism. They cannot give up the central policies that governments and corporations have followed for 30 years.

The point of this chapter is to explain this contradiction and why they can't give it up. This matters, because to halt global warming we will have to make them change, or move them aside.

The roots of neoliberalism

The reason the ruling classes of the world have turned to neoliberalism and market power is that they are desperate to raise profits.

They have been trying to change the world economy for 40 years. This is because in the late 1960s the rate of profit in industry fell by half across the developed world.[1] In Japan, for instance, industrial profits fell from 40 percent to 20 percent. In the US they fell from 25 percent to 13 percent and in Germany from 23 percent to 11 percent.[2] The same thing happened in Canada and across Western Europe. In the 40 years since the 1960s profits have recovered about half of the ground lost in the US, and very little in Europe and Japan.[3]

Why the Rich and Powerful Won't Act

To understand why this matters so much to corporate leaders, we have to start with why profits are so central to the global capitalist system.

Capitalism is new. Five hundred years ago power lay with big landlords, who lived off the work the serfs did on the land. The lords needed a steady income, but they did not have to invest money to make more money.

Then capitalism and industry grew together over the last 200 years. Machines came to dominate and corporations had to invest because they competed with each other. The winners were the companies that made things better and cheaper. But engineers were improving the machines all the time, so companies had to constantly buy newer, bigger and more expensive machines. The companies that spent the most grew and the ones that could not invest fell by the wayside. The money for all that investment came from profits. So the companies that made the most profits invested, survived, grew and came to dominate their industries.

That profit came from the people who worked for the companies. Their brains, muscles, hands, joints and sweat built and ran the machines. They carried, hauled, trucked and sold the goods. A smart or lucky company could get a jump on the others through new designs, products or production processes. But to keep the profits coming in, the company would have to spend everything it could on new investment. That meant holding down wages and making people work harder. The pressure was endless. Profits were the life blood of the corporations. The more profit they made, the more they could invest, and the more profits they could make.

Profits also mattered to nation states. This was because corporate executives in each country were always looking for ways to avoid the cold wind of competition. Wherever possible, they began to construct effective monopolies. In country after country "cartels" formed. These were alliances in each industry of the big oil companies, or the railroad companies, or the auto companies, or banks. The biggest corporations in any one industry tried to agree with each other to limit competition, to charge much the same prices, produce much the same goods and shut out new competitors. But even as cartels and monopolies formed in each country in the 19th century, competition moved to a global level between countries and empires. Economic competition between countries and empires led to war.

The strongest industrial powers produced the best arms. In 1800 the country with the best ships and cannons won. By 1860 the industries that made the most repeating rifles ruled the world. After that came machine guns, bridges, tanks, airplanes, bombs and finally nuclear weapons. In each era the great industrial powers became the great military powers.

The corporations that made more profits grew, survived and dominated. Nations that made more profit also grew, survived and dominated. This was why the fall in profits in the late 1960s mattered so much.

The crash hit profits from industry, not services. But industry accounted for most international trade and over half of total profits. Every country's competitive position in the world depended on its industry.

Investment in industry did not stop because of the fall in profits, but it slowed in the big G7 group of economies. In the USA, Canada, Britain, France, Germany, Italy and Japan, combined manufacturing output had been rising at 6.4 percent a year. Now it rose at 2.1 percent. But productivity was rising faster than this, so people began to lose their jobs as factories made more goods with fewer workers. The unemployed workers couldn't buy things, and so more factory workers lost their jobs. In 1973, and again in 1979, there were major recessions. Millions were out of work, and major corporations feared bankruptcy.[4]

The corporate executives didn't know why profits had collapsed, but they were feeling the pain. Business executives, bankers, politicians and mainstream economists didn't know how to *fix* the underlying problem, whatever it was. But they did know how to *cope*.

They had two strategies. One was global competition. The main form of this has been what is called "globalisation"—the attempt of the US government and corporations to dominate the world so that American corporations would get a bigger slice of global profits.

The other way of coping was to try to increase the share of the national income that went to corporations and the rich. This was done by cutting into the share of the national income that went to working people and their needs. That's what "neoliberalism" is. Neoliberalism is often described as the dominance of the market. That's not quite right. It's dominance through increasing inequality, so that more of the wealth goes to the corporations and less to workers.[5]

The corporations and politicians have spent the last 30 years reshaping everything they can in the world economy to make it more unequal. That does not mean everyone has got poorer. The world economy is still growing, though about half as fast as it did from 1945 to 1970. So there is more wealth to go round. But the rich have taken a bigger piece of the pie.

Why the Rich and Powerful Won't Act

How neoliberalism works

The ideas of neoliberalism and globalisation were most developed in the United States. But it was not just the American government putting them into operation. The corporations were having the same problem with profits in every country. All their governments welcomed the American model of neoliberalism.

Here's how neoliberalism worked in practice: The key strategies to restore profits were selling off public services ("privatisation"), cutting government spending, holding down wages, and scrapping government regulation.

After 1945 governments all over Western Europe nationalised—took over—many industries. In Britain, for instance, there was a national health service, free education and extensive public housing. The government owned the industries in water, sewage, natural gas, electricity, nuclear power, coal mining, steel, railways, buses, airports, airlines, post, telephones, prisons, the police, firefighting and the armed forces. The British government also owned some of the car factories, half the TV stations and all the radio stations. This was normal in Western Europe.

Then between 1979 and 1997 Conservative governments in Britain sold off coal, steel, natural gas, water, sewage, railways, buses, airports, airlines and telephones. Where these had made profits for the governments, they now boosted private profits.[6] Similar sell-offs took place all over the world.

Some industries and public services were politically very difficult to sell off—hospitals, schools, police, firefighters, the armed services and social services. Instead these services were honeycombed with private subcontractors. These subcontractors were not Fred the local plumber and four of his mates. They were almost all multinational corporations running thousands of contracts in countries all over the world.

From the point of view of the corporations, privatisation and subcontracting did three things at once. They increased private profit. They rehabilitated the idea that private companies can solve social problems. And they created a chance to weaken or stamp out union organisation.

This is the first reason neoliberal politicians—and that now means almost all mainstream politicians—don't want public action to stop climate change. People would go back to thinking that publicly owned companies can solve human problems.

The next way neoliberal policies tried to raise profits was by cutting down on taxes and public spending. Taxes ate into profits. The obvious first step, in almost every country, was to cut corporation tax. Next they cut the taxes paid by the rich. This also boosted profits, because the corporations did not have to pay as much to shareholders. The taxes paid by ordinary people also came out of wages. If the government cut ordinary taxes, the companies would not have to pay so much.

In order to cut taxes, governments had to hold down their spending. This meant holding down the wages of government employees, while making them teach more children in each classroom or care for more patients in each ward. And it meant providing worse services.[7]

Charges were introduced and increased. In the US financial aid for college was cut and tuition fees exploded, to $12,000 or more in a public university and up to $40,000 in a private one. The British, Australian and other governments eliminated student grants and introduced or raised tuition fees. In Africa the World Bank persuaded and bullied governments into charging for schools, clinics and hospitals. In the US health insurance costs ballooned, while Health Management Organisations cut services.

Much government expenditure also went on straight payments—pensions for the elderly, unemployment benefits, and welfare for the needy and disabled. Governments tried to cut all these benefits. Sometimes these cuts ran into too much opposition. Then the governments tried to means test benefits, throw people off benefit earlier and hold down the amount of benefit.[8]

This is another reason neoliberal politicians and corporations don't want the government to spend large amounts of money on stopping climate change. That would bring back the idea that government spending is the answer to human need.

Environmental and health and safety regulations also cost money and reduced profits. In Britain, the US and many other countries it was not politically possible to simply tear up the regulations. But the governments could avoid new laws and stop enforcing the old laws. Ronald Reagan simply appointed businessmen who would not enforce the rules to run the Environmental Protection Agency and other government bodies. In many countries governments also reduced the numbers of environmental and health and safety inspectors.

Governments all over the world also abolished many other controls on companies. Energy suppliers, stock markets, currency dealers, banks, airlines and many other industries were "deregulated". That also fed profits. This is why politicians and corporations are so reluctant to regulate industry and building construction in order to stop climate change.

Why the Rich and Powerful Won't Act

Finally, both governments and corporations did everything they could to make their employees work harder for less money. This meant holding down wages and cutting them wherever possible. It also meant attacking conditions. Tea breaks, smoke breaks, overtime rates and every kind of perk went under the hammer. The pace of work increased and the number of people doing the job fell. Frightened people worked through their lunch break and did an extra hour in the evening without pay.

To drive workers, employers had to make them afraid. Mass unemployment made that easier. People were afraid of losing their jobs and afraid of standing up for each other. But the employers also had to break the power of the unions. Sometimes they did this in great national strikes, where key unions were publicly humiliated— I outline a few examples below. But they also did it in office by office, school by school and building site by building site.

Alongside breaking unions was the relentless daily pressure from management to make people work harder. This had been common on factory assembly lines. It grew worse there, and spread through the public services and the professions. With that pressure went an epidemic of bullying.

The key word for how people feel under neoliberalism is stress. The pressure is constant. Some people are making better money than they did and some are not. But almost every kind of job, in almost every country in the world, is harder than it was 30 years ago. People feel pushed and helpless, in a world that has somehow spun out of their control. The values that dominate their jobs and their lives do not seem to them human values. And they are afraid—of losing their job, of what happens when their old mother gets sick, of losing their welfare payments, of being publicly humiliated at work. They are afraid of their own anger and of losing control themselves.

The battle of ideas

I have been writing as if neoliberalism is a conspiracy. And of course it is not quite like that. It was a process. Bankers, politicians, economists and political leaders tried different strategies and borrowed ideas and innovations from around the world.

Enforcing neoliberalism was never an easy project. It was always a struggle for the corporations and their politicians to make society more unequal. They had greater success in some countries—in

Britain, the US, Eastern Europe, China and Africa. They had less success in most of Western Europe. But nowhere did they win everything they wanted.

Every step of the way the corporations and those who supported them had to make people accept inequality. People had to be persuaded or forced to work harder, live under stress, have fewer public services and see the rich getting richer than them. By 2000, for instance, 79 percent of Americans made less than the average wage. Growing inequality was not in the interests of 79 percent of Americans. But somehow they had to be made to accept it.

The governments and corporations had two ways of pushing inequality through. One was a battle of ideas—they tried to make people accept that inequality was natural. The other was a battle of power—they had to show people there was no point in fighting back.

The battle of ideas was not easy. In America there was the memory of the 1960s. That was a time of great mass movements—civil rights, anti-war, women's liberation, gay liberation and more. All of them had been struggles for equality in one way or another. In Europe and elsewhere there was the welfare state. People had pensions, free healthcare, free schools, unemployment benefits and much more. They believed that was their right.

In the poor countries in Africa and Asia there were recent memories of great movements against colonialism. In the eyes of most people in those countries, these had been struggles for liberation and equality. In much of Latin America there was a deep tradition of both resistance to the US and fighting for social benefits.

All these ideas about inequality and a right to social care had to be broken if profits were to be forced up and kept up. People had to stop believing equality was possible. It's important to realise just how big these changes were, how much effort and struggle the rich and powerful put into them. This is one reason they dare not give them up now.

So from the late 1970s on there was a constant barrage of the idea that inequality was natural. This shift in ideas was strongest in the United States. Governments and companies funded research projects by academics that reached this conclusion. Television stations and magazines owned by large corporations also preached inequality. Books and movies argued for it, as did experts of all kinds. And these ideas from the United States spread round the world. This was partly because of the influence of American universities, American culture and American power. But it was also because the new ideas

Why the Rich and Powerful Won't Act

were what the rich and powerful elsewhere, in every country, wanted ordinary people to believe.

Myriad kinds of suffering were made the fault of individuals. Most kinds of suffering affect everyone, but they hit the poor far more often than the rich, and the average working person more often than the affluent professional. This is true of alcoholism, drug addiction, every form of mental illness, marriage breakup, loneliness, suicide, heart attacks, early death, panic attacks, committing murder, being murdered, child abuse, committing rape, being raped, committing robbery, being robbed, being imprisoned and domestic violence. Yet from the 1980s onwards all of these kinds of suffering have been relentlessly ascribed to people's genetic nature, from alcoholism to domestic violence. This implies that the reason black people and poor people suffer more is that they are genetically different—weaker, more violent, and less able to control themselves. It also implies that the reason ordinary people on average incomes suffer more in these ways than affluent professionals is that they too have worse genes.

The same myth that says social problems are genetic also includes a completely contradictory injunction—that the flawed individuals are supposed to solve their problems by pulling themselves together. There is no suggestion that if these people had damaged genes, maybe they should be helped. Instead *self-help* books, Hollywood movies and TV glorify winners and blame victims. If you feel like shit, you have a problem with self-esteem. No one ever says other people are treating you like shit, so you have a problem with the esteem of others. Every social problem becomes relentlessly personal. This is why the corporations and politicians who can't stop global warming now relentlessly push personal consumer decisions and lifestyle changes as the solution to climate change. They are accustomed to blaming people for their suffering and making them feel guilty.

The weakness of the left

One crucial victory for neoliberalism in the battle of ideas was the collapse of ideas and confidence on the left between 1975 and 2000. This collapse affected the far left, the communist, socialist and labour parties, and even liberal Democrats in the US. It also affected feminists and civil rights leaders in the US and radical nationalists in Asia, Africa and the Middle East.

There several reasons for this collapse. The first was that a whole generation of social democratic politicians and labour leaders had grown up during the long boom after 1940. These leaders had delivered the welfare state, and won much for union members. But they had done that as part of an expanding system. When profits crashed and the national economy was in trouble after 1973, these same leaders did not know what to do. They had always argued for a share of an expanding cake. Now the cake was shrinking, they didn't know what to say. When the company said they didn't have the money, the union negotiators believed them. They also believed there was no way unions could stand up to the world market.

Similarly, the politicians in the social democratic parties had indeed delivered real benefits when they were in government. Now they found themselves in charge of national economies in deep trouble. They believed both that they should defend working people and that they should defend their national interest. Now the two were in conflict. At first they were paralysed then did what business told them to. Their successful leaders became more and more right wing enforcers of the system. The majority of their activists quit or faded into inactivity. Membership fell. Working people, bitter at their governments, stayed at home on election day or voted for personalities.

In the United States in the 1980s a space suddenly opened for the top 10 percent of African-Americans and women earners. The incomes of the top tenth soared and they got jobs running parts of the system. They became press officers, generals, police chiefs, senior social workers and managers. They did the work of the system and began to believe in its values, turning their backs on struggles for civil rights and women's liberation. African-American politicians did this while the number of Americans in prison increased from 200,000 in 1970 to 2,000,000 adults and 700,000 children in the year 2000. Almost half those prisoners were working class African-Americans.[9]

In Asia and Africa the leaders of the movements that had led national liberation struggles now faced the same dilemmas. Almost all of them, in Egypt, India, South Africa, China and beyond, opted to enforce the world market for the good of the local rich. They became tyrants, or toys of the West, and their people became deeply disappointed and cynical.

The second reason for the collapse of ideas and confidence on the left was the fall of the Stalinist dictatorships in Eastern Europe and the Soviet Union in 1989. In Britain, as in many other countries, the left and centre of the Labour Party had traditionally been against dictatorship, but still thought that the East European regimes were

Why the Rich and Powerful Won't Act

in some sense socialist. Now the "communist" experiment had ended in tyranny and economic disaster, and workers too turned on the these regimes. That took the backbone, the belief, out of the Labour left and centre. This collapse went far beyond Britain. In Western Europe, Latin America, Africa and Asia the left suddenly did not know what to say to the neoliberals. For a time the idea of government intervention and nationalisation lost almost all credibility. In Britain the Labour Party swung behind Tony Blair and Gordon Brown, passionate supporters of the market.

Not everyone felt that way. I was a socialist who hated the Stalinist dictatorships. The workers' revolution of 1917 had long been destroyed by Stalin's counter-revolution and there had never been a workers' revolution in China, no matter how Mao's regime chose to describe itself. In fact both regimes were driven by the same logic of competitive accumulation that governs Western-style capitalism. They were the opposite of what I meant by socialism and I was glad to see them go.[10]

This is a particularly important point to make in a book on the environment. The Stalinist regimes in Russia and China wrought terrible environment destruction. Many environmentalists say quite rightly that they want no part of it.

Globally, the left, and many ordinary people, had thought Russian and Chinese "communism" proved there was an alternative to the market. With that gone, the corporations and those who spread their ideas now had a new weapon. Actual socialist ideas—about equality, dignity and democracy—were lost in the confusions about the Russian and Chinese tyrannies. This meant that the neoliberals did not have to glorify inequality as such. Most people would not have stood for that. And they don't say neoliberalism is about inequality. Instead they say it is about the "market". Their argument is that the market is the only way to do things. The market can make everyone rich. They say this even in places like Africa, where people are so obviously being made poor by the market. This glorification of the market is another reason that neoliberal politicians have insisted that all solutions to climate change must be market fixes.

Liars

Lies were also central to the battle of ideas.

Everyone has always known that politicians lie. But in the old days the British Labour Party, for instance, was able to give people part of what they needed. Labour had provided a national health service and a welfare state. In the US Franklin Roosevelt's New Deal gave people government pensions and Lyndon Johnson's War on Poverty gave free healthcare to the elderly. In the 1950s and 1960s liberal and socialist politicians did not have to fake concern for ordinary people. They served big business, and lied about that. But they also delivered social welfare reforms and they could tell the truth about that.

However, by the 1990s all the mainstream political parties supported neoliberalism. Clinton and the Democrats followed Reagan and the Republicans in the US. Blair and Labour followed Thatcher and the Conservatives in Britain. This was a global phenomenon. Mitterrand's Socialist government in France in the 1980s supported neoliberalism, as did Schröder's Social Democrats in Germany in the 1990s, and the ANC in South African after freedom was won. Now the Clintons and Blairs and the ANC did nothing for ordinary people. So they now had to lie about everything, all the time.[11]

To lie successfully, the politicians adopted the methods of advertising and called them spin. That made people even angrier. The American novelist David Foster Wallace wrote about why younger voters were so deeply cynical in the US 2000 election campaign:

> We've been lied to and lied to, and it hurts to be lied to. It's ultimately just about that complicated: it hurts. We learn this at like age four— it's grownups' first explanation to us of why it's bad to lie ("How would you like it if...?") And we keep learning, from hard experience, that getting lied to sucks—that it diminishes you, denies you respect for yourself, for the liar, for the world. Especially if the lies are chronic, systemic, if experience seems to teach that everything you're supposed to believe in is really just a game based on lies.[12]

Now politicians lie about global warming all the time. They tell us they have carbon trading schemes and they will cut emissions. At the same time they build more roads, more airports and more skyscrapers. Some of them do nothing. Some do a little something, but nowhere near enough, and they know that. Then they hold press conferences to tell us how much they care.

They lie about climate change because whenever any failure becomes public their first instinct is not to solve the problem. It is to hold a meeting with their advisers and decide what lie to tell, and to continue doing nothing about the underlying problem. They lie about climate change because they have become liars. And ordinary people feel they have no leaders to trust, and so no way to fight back.

Why the Rich and Powerful Won't Act

Smashing resistance

However, neoliberalism could not simply be won by a battle of ideas.

The neoliberal argument goes like this: *Inequality is natural. There is no alternative. We care about you. And you can't fight the market. If you try, you will be smashed.*

To prove that resistance was futile, actual resistance had to be taken on and beaten.[13] Three national set-piece defeats of the unions were key. In the US the air traffic controllers, all federal employees, went on strike in 1981. President Ronald Reagan fired every one of them on the third day of the strike and replaced them with barely trained scabs. Reagan got away with that—no planes ran into each other. The air traffic controllers lost and none of them ever got their jobs back. After that other American union activists were much more reluctant to take on the employers.[14]

In India, Mumbai (then called Bombay) was the industrial centre. A million and a quarter textile workers in that one city went out on strike in 1980 and stayed out for over a year. They went back defeated. In the next few years the companies closed most of the mills in Mumbai and reopened them in other Indian cities.[15]

In Britain everyone knew the mine workers were the strongest and most militant union. Prime minister Margaret Thatcher took on the miners in 1984. The strike lasted over a year, the miners lost and within seven years almost all the mines were closed. For years after that, every time anyone mentioned a strike at a union meeting anywhere in Britain, someone always said, "Look what they did to the miners".[16]

Political defeats for the opposition also strengthened the neoliberal project. In Nicaragua an urban insurrection in 1979 overthrew the hated US backed dictator, Somoza. The Sandinistas, the anti-Somoza guerrillas, were elected as the new government. The example resonated throughout Latin America. The US government stepped in, and American backed death squads exterminated mass guerrilla movements in neighbouring El Salvador and Guatemala. US funded Contra guerillas raided Nicaragua from surrounding countries. US sanctions crippled the Nicaraguan economy, until impoverished Nicaraguans gave in and voted the Sandinistas out.[17]

In China there was a mass uprising for democracy in 1989. It was led by the students in Tiananmen Square in Beijing. But the first time the troops were sent to smash the students, the workers in Beijing turned out in millions to talk to the soldiers and turn them back. Similar uprisings spread to every city in China.

The troops were sent back a second time. This time they shot the students. One photograph went round the world. For most people it symbolised Tiananmen Square. One lone young man stood defiantly in an empty street, defying the gun pointing at him from a giant tank. It was a symbol of courage, but also of the hopelessness of resistance. He was alone. No one thought he survived.

These were the big battles, the ones that set national and global examples. But even more important were the millions of small strikes, neighbourhood petitions and small personal acts of defiance that ended in tears.

Not all the small battles were lost, and not even all the big ones. In the USA gay men lived through AIDS and gained dignity, solidarity and respect. In South Africa the ANC and mass resistance in the cities finally broke apartheid. In France a general strike in the public sector won in 1995, and French workers have preserved much of their welfare state. Nor was the battle of ideas fully lost. Most ordinary people in the world continued to believe that governments should take care of the poor, the sick and the disabled. This is also why most people now believe that the governments *should* do something about global warming.

But in most places people came to believe the central right wing idea of the time that you can't fight back and win. You may not like the market and what is happening to your life, but there is no alternative. This was the great achievement of neoliberalism. This is what they stand to lose if government action stops global warming.

The weakness of neoliberalism

But neoliberalism still has two great weaknesses.

One is that it has not solved the underlying problem of profits. Neoliberalism was an attempt to cope, not to fix the system. The coping has worked. Almost everywhere society is more unequal. Even in countries where wages have increased, ordinary people are getting a smaller share of the national income. And the share that had come to them through the welfare state, pensions, benefits, health and education has fallen too. The share going to the corporations and the rich has risen.

But this has not touched the underlying problems that created the fall in profits in the first place. Different Marxist economists stress different elements that caused the fall. Some emphasise over-capacity, some global competition and some the effects of productivity and

Why the Rich and Powerful Won't Act

shrinking industrial jobs.[18] All these explanations point to reasons why industrial profits fell, and not other profits. And none of these problems has gone away. There is now more overproduction, more global competition and more productivity increases.

That means that even though corporations are getting more of the national income, profits have not risen to where they were in the 1960s. Of course the sums of profits are larger, but the economies have also grown. Profits are not a larger percentage of investment.

The partial recovery of profits in the US has been based on two things. Working Americans have been squeezed very hard. Their take-home wages per hour have not risen in 30 years, and American men now work almost 10 percent more hours than they used to.[19] The other reason is that US corporations have been the dominant power in the world. For all that, profits have only recovered by half, but this still places enormous pressure on Japan and Germany where profits have shown much less sign of recovery.[20]

Still the ruling class in every country cling to neoliberalism. They have no other solution to falling profits. Neoliberalism is still a way of holding the line, of stopping profits falling further. This is why corporations and governments hang onto neoliberal policies desperately. They dare not let up.

The other weakness of neoliberalism is that the assaults on people's incomes and lives have left behind oceans of bitterness. There are important places in the world where people have seen their incomes fall—in inner city USA, much of rural China, many parts of Latin America, Egypt, Iraq, Saudi Arabia and most of Africa. But far more general is the stress. Work is harder, life is more insecure, and the safety nets are frayed or gone. People lie awake at night and worry.

Anger and bitterness are everywhere. Mostly it is called apathy, which is the name for what anger mixed with helplessness feels like. In the 21st century that anger has begun to explode in the many new movements for social justice and against war. I come to those movements in a later chapter. In them lies our hope, not only for a more decent world, but also for the planet.

But that anger also meant that as the 21st century began the politicians and big business could not let up on neoliberalism. They had largely won the argument that there was no alternative. They were only coping with profits. They were sitting on a world of anger.

Sometimes the rulers, and their intellectual servants, say this explicitly. But all of them feel it in their guts. They have to hold the line.

Neoliberalism and Profits

The credit crisis

Then, from the autumn of 2007 on, it began to get more difficult for them to hold the line. The reason was the credit crisis, and the gradually spreading recession. This credit crisis, like neoliberalism, had its roots in the fall in industrial profits. The credit crisis came from the ways corporations and banks had been trying to *postpone* the problem. For since the 1970s corporations, banks and governments had been trying to make up for falling profits by speculation and borrowing. Those strategies began to come home to roost in 2007.

From the 1970s on, more and more investment went into speculative bubbles in real estate, shares, the dot.com boom, derivatives, securitized investment vehicles, commodity futures, even old masters of art and the like. So the housing market would boom not because houses were "worth" more, but because everyone involved was pretty sure the price of houses would keep rising. Then, at a certain point, the price of houses would crash.

This speculation did not solve the underlying problem of profits, however. Some people were making profits and others were losing, but no one was creating new value.

Banks, corporations and governments also had another reaction to the fall in profits. This was to increase the amount of debt and credit in the global system. When important companies looked like going bust, they were usually rescued. When the national economy was in trouble, the central bank would lower the rate of interest so it became very cheap to borrow money. On a personal level, this is why ordinary people in richer countries were able to run up such large credit card debts. But corporations and nations ran up debts on the same scale. Every time it has looked like the structure of debt will come crashing down, the central banks have simply made it even easier to borrow money. So the global amount of debt and credit has grown and grown.

Debt, speculation and neoliberalism all reinforced each other. Neoliberalism meant a bonfire of regulations that had limited speculation and debt. Increased credit meant more speculation was possible. Speculation thrived on increased credit. Everyone involved agreed the party would never end. And then it did. The whole structure of debt and speculation depended on all the banks and corporations agreeing not to call in their debts. In the autumn of 2007 that stopped. And then banks and companies had to pay their debts and didn't have the money to do so.

Why the Rich and Powerful Won't Act

But at root this was not because they had been foolish or reckless. It was because industrial profits had fallen and had not recovered.

The credit crisis has two main implications for climate change. First of all, it shows how governments were prepared to spend money. When the US economy looked like it was headed for a serious recession in 2001, the Federal Reserve Bank lowered the interest rate drastically. The intended result was that people borrowed money to buy new homes and new cars. That stimulated the economy. It also created an epidemic of McMansions and Sport Utility Vehicles (SUVs) that no one needed. That carried the economy along for a time and created the housing bubble that is now deflating.

But the government and the Federal Reserve could have done something quite different. They could have given people money, or interest free loans, to build low carbon houses, insulate old houses and build public transport. They were prepared to spend the money. They just weren't prepared to spend it on what human beings and the environment needed.

Even more important for climate change, the credit crisis has begun to discredit neoliberalism and the market. By the spring of 2008 there were opinion pieces every week in the *Financial Times*, the leading business paper, saying that when the crisis really hit ordinary people would no longer accept the arrogance of the market. Government intervention and social justice will be back in fashion, the *Financial Times* says.

This has already begun to open a new political space. In 2006 it still sounded visionary to demand millions of new climate jobs. By 2008 a conference in the U.S organised by the United Steelworkers union and the Sierra Club could seriously call for three million new green jobs in America. This was because both the Democrats and the Republicans were agreed that the economy needed massive stimulation. Neither party was prepared to do that by the government hiring people, much less hiring people to build wind turbines. But the idea made sense in a situation where something had to be done.

It is not possible, as I write this, to tell how long and deep the crisis and recession will be. But they will certainly deliver deep blows to the dominance of neoliberal ideas. That does not mean the corporations and politicians will give up such ideas and policies easily. After all, in an economic crisis they will want to squeeze working people even harder to keep profits up. But it does mean that ordinary people's minds, and hopes, will change.

CHAPTER 12

Corporate Power

THIS chapter is about the "carbon corporations"—oil, coal, gas
and motor vehicles. Here we are not talking about all corporations, or capitalism in general. We are talking about the particular
corporations that stand to lose most from changes to a new energy
economy. It is not that capitalists can't make money out of windmills. They did it in 17th century Holland and they can do it again.
But different companies will benefit, in different industries and different countries.

This is the way with technological change in capitalism. When
cars replaced railways, there were new families and companies,
Ford, General Motors and Volkswagen. IBM dominated computers
before software. Then suddenly it was Microsoft, and Bill Gates was
the richest person in the world.

Moreover, as I said in the introduction, the companies that stand
to lose most from global warming are also the ten largest corporations
in the world by sales in 2006. Here's a reminder of who they are:[21]

Ten largest global corporations by sales in 2006

1 Wal-Mart (US)
2 Exxon Mobil (US)
3 Shell (Netherlands/UK)
4 BP (UK)
5 General Motors (US)
6 Toyota (Japan)
7 Chevron (US)
8 DaimlerChrysler (Germany)
9 ConocoPhilips (US)
10 Total (France)

The top ten include the five largest corporations in the United
States, the most powerful country on earth.

The pattern of oil and car companies is repeated in country
after country. The top two corporations in Germany make cars,

Why the Rich and Powerful Won't Act

and so do the top two in Japan. The top corporation is an oil company in the UK, France, Brazil and Italy. In Russia the top four companies are oil and gas corporations. In India the top five are all oil companies.

This size means enormous economic and political power. Big corporations do not simply compete with thousands of other companies on a level playing field in an open market. In each industry only a few companies dominate the rest. In any one global industry there are still usually less than ten dominant companies. Their size and profits are the source of their political power, but their power is also one of the reasons for their size and profits.

Companies dominate because they are embedded in a history. They have a relationship with the government, the banks, the regulators, the suppliers, the sellers, the unions, the media and the other dominant companies in their industry. All of these relationships have been built up over time. They cannot simply be transferred to another industry.

In theory, a very brave corporate leadership could make that leap. In practice, it is so difficult and risky that it very rarely happens. A radical shift to a low carbon economy would reduce the market for oil, gas, coal and car companies to a tenth of what is now. That would mean corporate death.

The new wind and solar power corporations

Let's take a look at the statistics for the dominant corporations in wind and solar power. They show that three or four global companies dominate the market. Moreover, they are not the companies that dominated the old energy industry, and they are mostly based in different countries. This is immediately clear in a table of the top ten global solar PV cell producers in 2001 and 2004, measured in megawatts (MW) of electricity produced at peak times, when the sun is shining.

Leading global producers of solar power in MW of installed electricity

	2001	2004
Sharp (Japan)	75	324
Kyocera (Japan)	54	105
BP Solar (USA)	54	85
Mitsubishi (Japan)	14	75

Q-cells (Germany)	not available	75
Shell Solar (Germany)	39	72
Sanyo (Japan)	19	65
Schott Solar (Germany)	23	63
Isofoton (Spain)	18	53
Motech (Taiwan)	not available	35
TOTAL	296	952[22]

Several things are noticeable in these figures. First, the sales of the top ten more than tripled in three years. But the three firms that were utterly dominant in 2001 remained the top three in 2004. The electronics and microcomputer giant Sharp, in first place, dominates the others. Sharp accounts for more than a third of all the sales by the top ten, and more than three times the sales of Kyocera in second place.

Seven of these ten leading firms are from Japan and Germany, the two countries where the government first subsidised solar power. With the single exception of BP, they are not traditional energy companies. Electronics companies dominate this list because the technical problems of increasing output from PV cells made of silicon are very similar to the technical problems with silicon chips in computers and electronics.

The wind power market is even more concentrated. In 2005 ten corporations had 96 percent of the total global wind turbine sales:

Share of total global market for wind turbines, 2004

Vestas (Denmark)	34 percent
Gamesa (Spain)	17 percent
Enercon (Germany)	15 percent
General Electric (USA)	11 percent
Siemens (Germany)	6 percent
Suzlon (India)	4 percent
REpower (Germany)	3 percent
Mitsubishi (Japan)	2 percent
Ecotecnica (Spain)	2 percent
Nordex (Germany)	2 percent
All others	4 percent[23]

This list reflects the early support for wind power by the Danish and German governments. One company, Vestas, made over a

Why the Rich and Powerful Won't Act

third of all wind turbine sales. In 2005 they manufactured 100 percent of all wind turbines for use offshore everywhere in the world. Four companies controlled 77 percent of the onshore and offshore global market.

Three of these companies, with 19 percent of the market between them, were already major multinationals in other industries. General Electric makes power plants and electrical goods, and is the eleventh largest corporation in the world. Siemens (Germany) and Mitsubishi (Japan) are engineering multinationals. The other seven are wind companies.

The big four all produce for a world market. In 2005 Vestas made 25 percent of its sales in the USA. Other key markets were Canada, Spain, Germany, China and India. Gamesa made 44 percent of sales outside Spain and were opening plants in China and the US. Enercon had manufacturing plants in Germany, India, Brazil, Turkey and Portugal. General Electric manufactured turbines in the US, Spain and Germany, and was opening an assembly plant in China.

In short, both solar power and wind power are turning into classic markets dominated by a small number of multinationals. American companies only have a foothold. The old oil companies do not dominate and will find it hard to catch up. They may buy out major producers in the future—this is how Siemens has gained 6 per cent of the global wind market—but they are not doing so now.

SUVs

I turn now to the question of why the carbon multinationals can't change. Let's turn to a case study—the car industry and SUVs in the United States.[24] The US auto industry is particularly important because Americans account for 4 percent of the world's population, 23 percent of global CO_2 emissions, and a third of the CO_2 from all the vehicles on the planet.

The SUV saga is a story of how three of the largest corporations in the world met the threat of competition and change. They did it, not by changing their ways, but by looking to their government to support them against the world.

This is also a story about neoliberalism. It pushes flesh on the economic analysis in Chapter 11, and shows how Big Auto capitalised on the growing inequality in America to sell big cars to rich people. And it's a story about corporations and climate change.

Between 1980 and 2007 the US car industry in Detroit more than doubled the power of car engines. That extra power could have allowed them to halve carbon dioxide emissions. Instead they used it to build dangerous big cars which by 2000 could accelerate from 0 to 60 miles per hour as fast as a small car could in 1980. It's a story of the past, but also a warning for the future.

The story of SUVs begins in 1973. Three things happened that year. One was the beginning of the first major recession since the 1930s. The second was that the major oil producing countries in the Middle East and the global south began to flex their collective muscle in the Organisation of Petroleum Exporting Countries (OPEC). By holding down the supply, OPEC drove up the price of oil to $100 a barrel in today's prices.

The third key event in 1973 was a war between Israel and its Arab neighbors. Israel had won previous wars in 1948 and 1967 easily. In 1973 it was a close run thing. Most Arabs thought the Israeli victory would not have been possible without US support in money, weapons and intelligence. For a period of several months the main Arab oil producers punished the US by cutting off supplies.

This had little practical effect, because then, as now, most US oil came from non-Arab sources. But the oil boycott, combined with the shock of the oil price rise and the 1973 recession, frightened the American government and corporations. President Gerald Ford, with broad corporate support, decided the US government had to reduce US dependence on imported oil by cutting gas consumption at the pump. In 1975 Congress passed a federal speed limit of 55 miles per hour (mph) on all major highways. To everyone's surprise, "car addicted" Americans observed it.

Another new US law cut gas mileage in new cars by half. Gas mileage was calculated in terms of the miles per gallon (mpg) of all the new cars sold by any one company, averaged across the fleet. In 1975 this fleet average was 13 mpg. The new law said it had to be 18 mpg by 1978, and 27 mpg by 1984. Detroit managed the new limits by the early 1980s.[25]

This is important. People often doubt that corporations can make heavy cuts in carbon emissions just because the government tells them to. But the biggest and most important industry in the US cut gas mileage by more than half in less than nine years. Moreover, this was not done by carbon rationing or green taxes. The government just told the companies to do it, and they did it. That is how governments behave when they think something is important.

       Why the Rich and Powerful Won't Act

However, the story is not that simple or that happy. The new regulations presented the car companies with a major problem. Three companies dominated the American car industry. General Motors, Ford, and Chrysler were called the "Big Three" or simply "Detroit". By 1975 they were facing the same crisis in industrial profits as corporations in every other country around the world, and they were facing global competition.

Germany and Japan were forbidden to rearm after World War Two. So a much higher proportion of their national income was available for investment in industry—in new plants, machinery, research, and design. The result was that by 1975 Honda, Nissan, Toyota, and Volkswagen were making cars that cost less, lasted longer, and broke down less than the American Big Three. American consumers, not being stupid, began buying foreign cars.

The 1975 controls on gas mileage made the competitive position worse for the Big Three. The foreign companies had begun by specialising in small cars because back then Japanese and German consumers were poorer than Americans. Japanese and German cars also got better gas mileage, because they were better engineered. And Detroit could not get their gas mileage down fast enough simply by changing the engines. The government regulations required each company to reduce the average mileage of *all* the cars they sold. So the easy way for the Big Three to meet the target fast was to sell more small cars. But too many small cars were already chasing too few buyers.

In the 1970s and 1980s new producers were building plants all over the world. By the 1990s these factories could make three small cars for every two that were sold in the world. Companies everywhere dropped their prices on small cars. It was a hard market in which to make money, and the Big Three were bad at making small cars anyway. Ford nearly went bankrupt during the second recession of 1979-82. Chrysler did go broke, and had to rely on a massive government rescue loan.

What American car companies were good at was big gas guzzlers. To get round the gas mileage rules, they invented a new kind of vehicle—the Sports Utility Vehicle. To make an SUV, the companies took the basic design and frame of a pickup truck. They put the body of a big car on top and added four-wheel drive. The trick was that the government agreed to class these SUVs as "light trucks" instead of as "big cars".

Light trucks came under a different set of government regulations. They had to run on 20 mpg, not 27 mpg. The biggest light trucks and SUVs had no restriction on gas mileage at all. The safety rules were more lax. Most important, light trucks were subject to a 25 percent import tax, while cars came in without an import tax. That extra quarter on the price kept the Japanese and German companies out of competition with American SUVs.

The Big Three had now found a niche where they had their monopoly back. While there was a glut of capacity in small car factories, there was a global shortage of capacity in light truck plants. Detroit could charge what they liked.

SUVs and political power

The new SUV trick depended on getting the federal government to do what the Big Three wanted. Big Auto had, and still has, several kinds of power. First, in the US one advertising dollar in every seven comes from the car industry. This is concentrated on magazines and television. Car industry leaders meet regularly with editors and publishers. They don't have to threaten to withdraw advertising from any outlet that strays out of line. Everyone knows they will just do it, quietly.

Secondly, they have access to politicians. This is not mainly a matter of buying Congress. The Big Three only contribute small amounts to political campaigns. But the politicians already believe the car companies are important. For many years, from the 1950s to the 1980s, it was routine for economists and business journalists to judge the state of a country's economic health by just one measure—new car sales. No American president since the 1920s has wanted to preside over the destruction of the country's largest industry. In 2002, for instance, GM had seven times the sales of Microsoft. GM and Ford together sold more than the entire American airline industry. The combined sales of GM and Ford were also bigger than the whole federal defence budget.

Thirdly, the Big Three often lobby government alongside the union. The United Auto Workers (UAW) is the traditional heart of union militancy in the US and has always worked closely with liberal Democrats. Even in 2002 the union still had 800,000 members, plus 1,700,000 retired members and their spouses on union pensions. The UAW gets out the vote, and its members are concentrated in the swing states of Michigan and Ohio. Ohio gave George Bush the margin of victory in 2000 and 2004.

Even with all those votes, when the UAW opposes the corporations, it does not have that much influence with politicians. The

Why the Rich and Powerful Won't Act

union campaigned hard against NAFTA (the North American Free Trade Agreement) and lost. But when the UAW lobbies alongside the corporations in Washington, the union delivers the liberal Democrats and the Big Three deliver the conservative Democrats and the Republicans.

When the Big Three wanted to fight the new mileage per gallon limits in 1975, the union refused to support them. But from 1980 onwards the union supported the corporations over SUVs for several reasons. American unions had stopped growing. The heart, the crusade, had gone out of them under the neoliberal assault. Foreign companies were opening up new plants all over the country and keeping them non-union. The UAW's membership was concentrated in American-owned plants making big cars and light trucks. The union decided their interests lay with the American car corporations.

Union members are now paying for that mistake. In the short run the strategy worked. Business and union lobbying together won a whole series of exemptions for light trucks and SUVs. Foreign companies had been importing light truck kits and assembling them in the US to avoid the 25 percent tax. That loophole was closed. Business men and women found it hard to claim tax exemptions for cars, but easy for light trucks. 1990 saw a new luxury tax on cars over $30,000, but not light trucks. The Clean Air Act of 1990 reduced emissions further for cars, but much less for light trucks. Safety rules allowed light trucks a longer stopping distance. They could have much cheaper and weaker brakes. The roofs of light trucks were exempt from federal strength standards, and inferior tires were legal.

SUVs were also expensive, really expensive. The Big Three often cleared less than $1,000 profit on a small car. They could make $30,000 profit out of the $50,000 price of an SUV. By the end of the 1990s, SUVs still made up only 17 percent of new car registrations in the US, but they provided most of the profits for the Big Three.

In 1998 one Ford SUV plant, the Michigan Truck Plant in the Detroit suburb of Wayne, was the most "profitable factory in any industry anywhere in the world... There were fewer than a hundred companies in the world that earned more than that single factory did in 1998".[26] The Michigan Truck Plant made a third of Ford's total profits. The value of its sales was bigger than Nike, and almost as big as the global sales of McDonald's. The profits were nearly as big as the total at Microsoft.

One factory.

By the 1998 sales of SUVs had brought GM and Ford back to healthy profits. Chrysler, the weakest of the Big Three, had been bought out by the German firm Daimler. That year two thirds of the new SUVs in the world were bought in the US.

SUVs and inequality

The SUV bonanza relied on political influence. It was a response to competition. But the reaction was not to compete in that market; it was to reshape and control the market through political clout. This is normal. It was not some strange departure from perfect market conditions but is how real corporations and real markets actually work.

But the SUV bonanza was also an adaptation to what neoliberalism was doing to American society. Inequality was growing. The biggest gap was not between rich and poor, but between the top 10 percent and the average people in the middle. And the people in that top tenth were learning to despise those beneath them, but also to fear them.

SUVs catered to the growing market among that richest tenth of the population. In order to sell, the design and advertising pitch was adapted to fit the emotional needs of these newly affluent people. SUVs were supposed to be safe for the driver and dangerous to other drivers. They expressed both the contempt and fear of the powerful for those beneath them—literally beneath them, lower on the road.

Households making over $50,000 dollars a year bought two thirds of new cars. Two thirds of the rest were bought by their children or by retired people who had also been making that kind of money. In total, affluent families were buying eight out of nine of all new cars, including the small ones. And it was the richer among the affluent who were buying the SUVs.

Only some of the rich, though—those whose characters best expressed the new inequality and the growing bullying at work. The marketing people from the car companies did considerable research to find out who was buying SUVs. It turned out that SUV buyers were likely to be baby boomers, professionals in their forties. In the 1960s they smoked some dope and had alternative opinions. Now they were tied to jobs and big organisations, but like to think of themselves as free spirits tearing round the countryside. They tended "to lack self-confidence in their driving skills".

Why the Rich and Powerful Won't Act

They were "frequently married with children, but often uncomfortable with both".[27] They were more selfish, less likely to care about other people, and they worried a lot about how they looked to other people. They liked control. Most of them lived in the suburbs of large metropolitan areas, especially Greater New York.

Clotaire Rapaille, a French medical anthropologist, was a key marketer for Chrysler in the 1990s. Keith Bradsher, Detroit correspondent of the *New York Times* and author of a wonderful history of the SUV, *High and Mighty*, interviewed Rapaille:

With the detachment of a foreigner, Rapaille sees Americans as increasingly fearful of crime. He acknowledges that this fear is irrational and completely ignores statistics... The fear is most intense among today's teenagers... "There is so much emphasis on violence— the war is every day, everywhere," he said in an interview two weeks before the terrorist attacks of 11 September 2001. The response of teens, he added, is that, "They want to give the message, 'I want to be able to destroy, I want to be able to fight back, don't mess with me'."

While teens do not buy many SUVs, youth culture nevertheless tends to shape the attitudes of broad segments of American society... People buy SUVs, [Rapaille] tells auto executives, because they are trying to look as menacing as possible to allay their fears of crime and other violence... "I usually say, 'If you put a machine gun on top of them, you will sell them better... Even going to the supermarket, you have to be ready to fight'..."

SUV buyers want to be able to take on street gangs with their vehicles and run them down, he said, while hastening to add that television commercials showing this would be inappropriate.[28]

What the SUV ads show instead is great big scary vehicles coming right at you, fast. The front lights and bumpers are explicitly designed to look like the eyes and jaws of a monster. A big selling point is that in a collision an SUV will crush a small car. This is true. In a frontal crash the higher SUV simply opens the top of the other car like a tin can. They are even more lethal in side collisions.

Other risks, however, are worse for SUV drivers and passengers. They are far more likely to roll over, because the higher a car in proportion to its width, the more unstable it is. SUVs are also more likely to swerve off the road, they are harder to control, and the cheap brakes, bad tires and fragile roofs are dangerous.

The SUV crash

In the 1990s the American government stopped worrying about oil supplies. The Middle East was under control, and the price of oil had been low for years. In 1990 the 55-mile an hour speed limit was abolished. Under both Clinton and Bush, attempts to tighten gas mileage laws got nowhere in Congress.

The SUV boom depended on class arrogance, class fear and neoliberal inequality. It also showed that the Big Three, like most corporations, did not react to new circumstances with fundamental change. Instead they continued to exploit the skills they had, and the political and economic niche they already owned. And they did little to develop well built, cheap small cars. Then they got hit by 9/11, Afghanistan, Iraq and climate change.

The price of oil rose after 2001 for four different reasons. The US invasion cut Iraqi oil production. Global demand rose at the top of a boom. Peak oil restricted supply, and nervousness about future Middle Eastern supplies pushed the markets higher. So Americans had to pay a lot more at the pumps. They reacted the same way they had in the 1970s—they bought small cars.

The environmental movement was also finally burying the SUV. It had taken a long time. Through the 1980s and 90s the environmental organisations had done almost nothing on the issue. One reason was that green organisations were staffed by single people from well to do families who wanted to campaign about elephants, whales, coral reefs, rainforests and polar bears. Of the thousands of full time professionals in environmental organisations, only eight were specialists in vehicles, and all eight lived in Washington, DC, or Berkeley, California, not Detroit.[29] Another reason for silence was probably that the environmental organisations relied for their income on contributions from the same affluent baby boomers who were buying SUVs.

Once environmental organisations began to really take climate change on board, the affluent followed. SUVs had been particularly big in Hollywood, where many people cared a lot how they looked. Now, for the same reason, the stars have led the rush to hybrid vehicles like the Prius. Most people couldn't afford such a car. Indeed, most people couldn't afford a new car at all. But among those who could, SUV sales were crashing. By 2006 Ford and GM were going down with them.

Toyota made the Prius. German, Japanese, French and Korean companies made the best cheap small cars. Ford's losses in 2006

Why the Rich and Powerful Won't Act

were $14.6 billion. General Motors was cutting a third of its workforce and looking for a foreign buyer. Ford planned to lay off half its workers. Daimler sold Chrysler to cut its losses.

Meanwhile, the United Auto Workers had made a terrible miscalculation in following the Big Three. Both Ford and GM were on the edge of bankruptcy. In America that didn't necessarily mean the owners lost control. Instead the company would go into "Chapter 11" bankruptcy, supervised by appointed accountants. It could continue to trade for a time, while the new bosses worked out a deal with the unions to put the company back on its feet under the old owners. If the union did not agree that deal, then the company would close. That was the threat.

Over the years the members of the UAW had won themselves the best pensions and medical benefits of any blue collar workers in the country. That meant the car companies had a large overhang of debt to their former workers. Under "Chapter 11" bankruptcy, the accountants would try to persuade the existing workers that they could only save their jobs by sacrificing the terms of other people's pensions.

The whole story of SUVs also stands as a warning to anyone who thinks that the old industries of cars and oil will adapt to a new economy of alternative energy. Instead they see adaptation to climate change as corporate death. They will make whatever green noises they are forced to and no more. The SUV story also stands as a warning to unions. When action against global warming threatens jobs, there is always a temptation for unions to ally with the corporations to defend the industry. In the end, that's a mistake. It won't save the jobs.

Between 1980 and 2007 the mileage per gallon rules did not change, but engineering progress did not stop. Relentless development produced far more powerful engines. That extra power could have reduced gas mileage by half again, to a quarter of what it was in 1973. Instead it was used to make big fast cars that threatened the planet and sold well *because* they could hurt people.

Technology is not the problem. It's what corporations do with the technology.

Big Oil

My second example in this chapter is the oil industry.[30] Many companies in many countries have combined into cartels and monopolies that control one line of production. But for 80 years a global cartel has controlled the oil industry around the planet, and that cartel now includes six of the ten largest corporations on earth.

The cartel was founded under the leadership of the Rockefeller family. John D Rockefeller, the founder, built his Standard Oil empire in the 1870s by taking over or driving out competition. Back then most American oil fields were in Pennsylvania and Ohio, and railroads were the only efficient way of moving oil. Rockefeller signed sweetheart deals with the railroads. Indeed, he pioneered a new kind of deal. Under it, a cartel headed by Rockefeller paid $2 a barrel to transport their oil, but the railroad then paid them a 50 cent rebate. The cartel's competitors had to pay the same $2, but the railroad then gave a rebate of 50 cents on each of those barrels to the Rockefeller cartel. This meant that even if the Rockefeller cartel shipped no oil, they would still make a steady profit. And it meant their competitors could not compete.

Tricks like that required, of course, that Rockefeller owned politicians and judges. The oil, railroad and other cartels corrupted American politics on a massive scale. And they led, in the end, to a populist movement determined to cut the monopolies down to size. With the support of President Theodore Roosevelt, the federal government eventually took Standard Oil to court, and won a massive anti-trust suit in 1911. Standard Oil was forced to split up into several companies that in theory competed with each other. In practice, the Rockefeller family was the largest shareholder in all of them, and they continued to act together, if more discreetly. After some remergers those oil companies have become, among others, Exxon Mobil, Chevron and Conoco Phillips.

In 1928 the Rockefeller method went global. In September of that year the executives of the three largest oil companies in the world—Exxon, Shell and BP—met at Achnacarry Castle, a 50,000 acre estate on the West Coast of Scotland. The journalist Linda McQuaig, in her angry and delightful book *It's the Crude, Dude*, describes the result of that meeting:

> The written document, dated 17 September 1928, was an agreement among the three dominant companies not to compete with each other, but instead to set quotas in order to maintain their existing market shares… The broad agreement of principles was followed by three much more specific agreements, drawn up over the next six years, [and] four more big players—Texaco, Gulf, Mobil and Atlantic—were brought in…

Why the Rich and Powerful Won't Act

The agreements are astonishing in their detail. Every aspect of maintaining market share was set out with great specificity: how quotas could be revised under different circumstances, and what should happen when quotas were exceeded or when they were underused... There were even rules to be followed in the area of corporate advertising. Types of advertising that were to be "eliminated" or "reduced" included road signs, billboards, newspaper ads, signs at dealers' garages, and novelties such as cigarette lighters. More basically, it was specified how prices were to be set (by mutual agreement) and how outside competitors were to be dealt with (preferably by acquiring them).

All this was overseen by a full-time secretariat with offices in New York and London, with the mechanics of the day-to-day implementation being left to what are unabashedly described in the local documents as "local cartels"... The extent of their activities is captured by the fact (uncovered by a Swedish parliamentary investigation) that in 1937 the oil companies' local Swedish cartel held a total of 55 meetings at which 897 subjects were discussed. (The following year, 656 subjects at 49 meetings, and the year after that, 776 subjects at 51 meetings.)[31]

If that was the Swedish branch of the cartel, think what was happening in the US or Britain.

The oil companies later claimed that this agreement was wound up in the 1930s, but it was certainly still operating in the 1940s and in many ways right up to the early 1970s. One startling piece of evidence for this is that world oil supply grew at "almost exactly 9.55 percent" every single year from 1950 to 1972. In no year was it less than 9.45 percent or more than 9.65 percent. When you consider how much local geology, accident and competition might have influenced the supply, there is only one possible explanation. The major oil companies had decided production would grow by 9.55 percent, and that is what happened.[32]

The major oil companies were controlling the supply of oil. Had supply and demand ruled the market, oil would have been very cheap and the oil companies would have made little profit. So they restricted the supply of oil. But they did not restrict it that much. Had they simply followed their own interests, oil would have been far more expensive than it was. The oil companies were also acting on behalf of industrial capitalism as a whole, and enjoyed the active support of Washington and the CIA because they were willing to keep the price of oil low enough not to worry Western industry.

From 1945 on, though, it was clear that most of the world's oil was in the Middle East—in Saudi Arabia, Iran, Iraq, Kuwait, Libya and the Gulf Emirates. Without oil these were poor countries. It was in the interests of the Arab and Iranian people to control their own oil and get the revenues. It was also in their interests to restrict supply and keep the price high. To deprive the local people of the wealth they sat on, brutal dictatorships allied to the United States and friendly to the oil companies were necessary. This is why Middle Eastern politics has been so cruel, and why the US government is so widely hated across the region.

The crucial player with the most oil has been Saudi Arabia. The kingdom, a dictatorship, has been a close ally of the United States ever since President Franklin Roosevelt visited it secretly just before his death in 1945. Until the 1970s all Saudi oil was controlled by Aramco, an American oil company in which Exxon was the major shareholder. Until 2007, whenever it looked like the price of oil was going too high, the Saudi government obeyed American requests to turn on the taps and get the price back down.

Throughout these years, from 1928 to 1972, the major oil companies kept out competitors and broke unruly governments. In 1953 an elected government in Iran took their oil fields away from the Anglo-Iranian Oil Company (now BP). Though this was a British company, it was the CIA which organised the coup that deposed the elected government and installed Reza Shah's dictatorship.

But in 1969 the oil majors faced a bitter blow. That year the Arab nationalist Colonel Gaddafi deposed the King of Libya. Gaddafi nationalised Libyan oil, and John Paul Getty's small Occidental Oil company broke ranks with the majors and agreed to market Libyan oil. In 1972 Saddam Hussein's government nationalised Iraqi oil. Even American allies got in on the act—Saudi Arabia negotiated a 51 percent share in their own oil, and the Kuwaiti and Gulf kingdoms took over their own fields as well. The major Third World oil countries, joined together in OPEC, then began to try and control the price of oil themselves. And in 1979 the Iranian Revolution finally took back Iranian oil.

Big Oil never forgave Gaddafi, Saddam and the ayatollahs. But the oil majors survived and thrived. They continued to control the shipment, distribution and much of the refining of oil. OPEC controlled much of production. This was made easy for them because they were basically taking over part of the existing monopoly cartel of Big Oil. But while OPEC were able to increase the price of oil briefly in the 1970s, the Saudi Arabian alliance with Washington

Why the Rich and Powerful Won't Act

continued, and oil majors, to this day, have maintained their position in the front rank of corporate power.

After 2000 it gradually became clear to the oil majors that the world was reaching Peak Oil, and the remaining reserves would become even more valuable. Nor had the oil companies forgotten the historic blow of nationalisation. The American invasion of Iraq in 2003 had several causes. But one of them was the longing of the oil companies to take back the oil fields in Iran and Iraq, the second and third largest fields in the world. This is what Cheney and Bush meant when they spoke of "bringing democracy" to Iraq— not elections, but free enterprise, and above all selling off the oil fields. This is why the control of Basra, in southern Iraq, remains critical to the Bush administration. That's where most of the oil is. And that's why Cheney and the oil companies he represents still dream of attacking Iran.

As in Pennsylvania and Ohio in the 1870s, so in Iraq and the world in the 2000s. Oil companies can only function by influencing and controlling governments. The global oil cartel is coming under pressure from Chinese, Russian and Venezuelan competitors. But the majors still control the largest oil fields and the world market. And that control is political. It's not because they are good geolo gists or clever investors.

That's why Shell and ExxonMobil can't make the leap to becoming wind and solar power companies. They can buy up and run small subsidiaries. But their global position depends on an established network of politicians, governments and armies. They cannot transfer that to wind. They have fought action to halt climate change every step of the way. Until the last few years they were quite open about this. Now they are compelled to be more sneaky, to work the back rooms of power. But they will not go away.

CHAPTER 13

Competition and Growth

W E have looked at two main reasons the rich and powerful are now reluctant to act on climate change—neoliberalism and the power of the carbon corporations. Neither of these are problems with the capitalist system as such. In other times and places capitalism has coexisted with far more state intervention than under neoliberalism. And while a clean energy economy threatens carbon corporations, other corporations could make massive profits.

This is all in theory. In practice, actually existing capitalism may be able to adapt and change. Or it may not. After all, the carbon corporations are powerful. And neoliberalism is not a hobby for governments and corporations. Profits are at the heart of the system. From the 1940s to the 1960s policies of public investment seemed to stimulate profits. Since then profits have been in trouble and the rate of growth in the world economy has slowed to half what it was. Corporations and established governments may be too afraid to contemplate a return to the old ways.

However, two more aspects of capitalism make action against climate change difficult. One is global competition. The other is relentless growth. Both are constant, central parts of the dynamic of the system.

I will take global competition first. Different corporations within one country are constantly in competition with each other. At the same time, however, the corporate leaders are trying to avoid competition within the country by forming cartels of the leading companies in each industry. In each country these cartels or monopolies work through the national government. That government defends and protects the interests of all its corporations against the other corporations in the world.

This has produced multinational corporations that pursue two strategies at once. They try to build manufacturing plants and increase sales in countries all over the world. That is why they are "multinationals". But they also depend on one home state to support them. So General Motors is both American and global, Toyota is both Japanese and global, and Total Oil is both French and global. So even as multinationals become increasingly powerful, national alliances remain central to corporate strategies.

This global competition has two sides. One is that each corporation is still attempting to make larger profits than its global competitors in order to invest more, capture more of the market, make larger profits, invest more and so on. Here, as with national competition, the corporations that do not invest shrink and go bankrupt or are bought out.

The other side of this global competition is political competition between different states. If one national government and its corporations are stronger, they can force others to do their bidding. This competition is not simply a matter of economic and global power. Behind that is the threat of war.

This global competition is a serious obstacle to action against global warming. The national capitalisms in each of the major powers have to make profits and therefore must constantly invest and grow. Action on climate change may create jobs for ordinary people, but it is a cost to each national capitalism. Each national alliance of corporations tries to make sure that the other national alliances pay more of the cost. The difficulty is that any solution to climate change has to be global. In the end, it has to involve agreement between those competing powers.

This problem of competition for profit can sometimes be solved at a national level. In 1863 in Britain pottery owners sent a petition to parliament supporting a law to reduce working hours. It said that the pottery owners wanted to better the conditions of their workers, but if any of them did so they would lose out to their competitors. So they wanted parliament to pass a law, and then they would all have to obey it equally.

Many countries have since passed laws limiting the working day. It is possible—not easy, but possible—to win these sorts of reforms within one country. But it is much harder to win them on a global level, where there is no common legal system to enforce them.

Globalisation now

The ideas of neoliberal globalisation were developed in the United States from the late 1970s on. If neoliberalism was a project for increasing inequality between the rich and ordinary people in every country, globalisation was the project for increasing the share of global profits going to American corporations.

From the point of view of US corporations, they would solve their problem with industrial profits in three ways. They would take from working people in the United States. They would win a greater slice of global profits for themselves. And they would force working people all over the world to give up more to the corporations, which would make it easier for American corporations to make profits around the world.

From about 1979 to 2003 the corporations and political leaders of the other major industrial powers had a complex reaction to neoliberal globalisation. The leaders of Japan, France, Germany, Italy and Russia were enthusiastic about the neoliberal project to take from their own working people. After all, all the major powers had the same problem with industrial profits as the United States. Neoliberalism looked like the answer.

But they didn't like the part about American corporations dominating their corporations in their own countries and on a global level. Yet neoliberal globalisation was presented to the world as a package, in the WTO, in the IMF, in the World Bank and in the realm of ideas. The other world leaders wanted part of that package, but not all of it. This is why the governments of Germany, France, Russia and the other major powers have gone back and forth in their relationship to American power. As American global power weakens, the other major powers now see increasing opportunities to stand up for their own interests.

Britain

There were exceptions among the major industrial powers. Canada and the US effectively form one economy, and Canada follows the American lead. Britain too is closely linked to the US. From 1980 to 2002 the US accounted for about half of all direct foreign investment coming into Britain, and a third of British overseas investment went to the United States. That made Britain the largest investor in the US, and Britain invested more there than in any other country. In 2006 Britain was the leading recipient of foreign direct investment in the world. Britain had more than three times the incoming investment of the next ranked country, China.[33]

Why the Rich and Powerful Won't Act

Moreover, the third largest corporation in the world in 2006 was Shell Oil, 40 percent British owned. The fourth largest corporation in the world was British Petroleum. The four largest oil companies in the world (Exxon Mobil, Shell, BP and Chevron) all depended on American military power to defend their interests in the Middle East.

In addition, British manufacturing industry was declining. The British ruling class reacted by building London up as a global financial centre, the place where American banks and corporations did business with European and Middle Eastern capital. By 2007 Britain was still the sixth largest industrial power in the world, but more people in Britain were working in banks and financial services than in industry.

The implications for climate change

As we will see in the next chapter, this pattern of global competition is reflected in international negotiations on climate change. These have pitted the United States against Europe, with Canada and Britain in between.

Until the 1970s the United States domestic oil industry was the largest in the world. So the US developed in a way that was wasteful of oil and energy. This is why the US now has CO_2 emissions of 20.2 tons per person. Western Europe, without its own oil reserves, was more careful of energy and now has CO_2 emissions of 8.8 tons per person.

Moreover, the US government has been committed to the interests of its three leading oil companies and three leading car companies. The power of these corporations, and their high level of CO_2 emissions, meant that it was more difficult for the American ruling class, and American corporations, to join any international agreement to limit emissions. Germany, France and Japan, on the other hand, had a competitive interest in forcing American industry to cut emissions. This meant they didn't want deep cuts that would affect their own industry, but they did want cuts that would hurt American corporations.

This competition structured all international climate negotiations from 1992 to 2007. The US government, under the elder Bush, the male Clinton and the younger Bush, did everything they could to make any international agreement as weak as possible. Then the younger Bush refused to sign Kyoto and looked for ways to sabotage any subsequent agreement.

India and China

The policies of neoliberal globalisation increased American domination of the world economy and increased inequality in countries around the world.

India and China, however, stand out as partial exceptions among the poor countries. The Chinese dictatorship and elected Indian governments of all political stripes have embraced neoliberalism. They have cut public services, privatised and massively increased inequality. But they have largely been able to resist American power. The governments of both India and China retained more control of customs duties, movements of money, foreign investment and economic policy than did most other governments. This, combined with very large internal markets, has made possible very high rates of growth.

This is important. The two poor countries most resistant to American power are the two that have grown the most. In both countries the ruling class has pursued a strategy of global competition through low wages and heavy investment.

All this has implications for climate change. Chinese and Indian growth is predicated on low wages, limited public services and global competition. This means that, even as both economies grow, the ruling class is loath to allow the sort of public programmes for clean electricity and energy efficiency that would challenge neoliberalism.

The Indian and Chinese ruling classes are also enthusiastic participants in what is now a global "race to the bottom". This involves all the major industrial powers. India and China are the leading low wage competitors. The American corporations have managed to hold down the wages of most working Americans, so that they have hardly grown in 40 years, and have cut the services available to ordinary people. This puts particular pressure on the ruling classes of the European Union. In Western Europe wages have risen more, and much of the welfare state remains. The EU, and the governments of France, Germany, Italy and the rest now feel under constant pressure to cut welfare spending, hold down wages and pensions, raise the retirement age and abolish laws protecting labour rights. Despite fierce opposition from their populations, the European governments keep coming back with more and more proposals for "reform".

This global competition makes it difficult for any ruling class to contemplate cutting CO_2 emissions on its own. Governments can

Why the Rich and Powerful Won't Act

agree in principle that it is essential to stop global warming. But when it comes to actual international negotiations, their commitment is to making sure that their competitors spend more and they spend less.

Growth

Global competition is a constant feature of capitalism. So is growth. The real problem here will be growth in the rich countries, not the poor. As we saw earlier, capitalism grows through competition. But that competition takes the form of investment to produce more and more goods. That production has so far meant constant rises in carbon dioxide emissions. If we don't take action against climate change, emissions will continue to rise.

However, all is not doom and gloom here. Capitalism needs growth, but if we take action against climate change that growth will be in sectors like clean electricity, efficient buildings and public transport. In that case we can have economic growth in the gross national product *and* falling emissions.

This will only be true for a time. In the long term the dynamic of capitalist growth will reach limits set by climate change. But if we act now, that won't be a problem for at least 20 years. And that's the crucial period for now, the period we must act in. So for the moment the planet can live with growth.

I will return to the long term problem in the last chapter, when I look at just how far we will have to change the world to save the planet.

The argument so far

I will now summarise my argument so far:

We have to cut carbon dioxide emissions by about 80 percent per person in the rich countries within 30 years to avoid abrupt climate change. Cuts in emissions on that scale are technologically possible now. But they will require massive public works and government intervention and regulation on a global scale.

The rich and powerful don't want to take that kind of action because they want to protect profits and neoliberal ideas. Some powerful carbon corporations will also be threatened by any serious action. And serious global action is difficult because the logic of capitalism drives countries to compete.

All this means effective action against climate change is possible, but the rich and powerful have strong reasons not to do it. That means we have to develop a global mass movement to make them act, or replace them with people who will act.

So the next chapters turn to climate politics. I look at what the ruling classes have actually done, and trace the effects of neoliberalism and competition and the strategies of carbon corporations. But I also look at the global movement for action so far. The aim is to learn from our strengths and move beyond our weaknesses.

Why the Rich and Powerful Won't Act

PART FOUR

CLIMATE
POLITICS

CHAPTER 14

The Road to Kyoto

THESE are the best of times and the worst of times. Suddenly climate change is everywhere in the media. Every great polluting corporation runs ads about how green they are. Leading politicians of the centre and the right rush to claim the issue as their own. George W Bush was mocked around the world, and finally had to admit that, yes, it is getting hotter.

But all the politicians who talk of action also talk of personal choices and carbon markets. Global carbon emissions keep rising. Governments set carbon targets that are not enforced, and then encourage corporations to build more roads, more cars, more airports, more planes and more leaking houses. The governments of America, China, India and Japan are building an alliance to do nothing.

People talk at work, on buses, in the pub, about how hot it is, and how something must be done. Global rock concerts spread the message.

Demonstrators all over the world have begun to demand action over climate change. Most of those protests have been small. But 150,000 protested in over 800 US cities and towns in April 2007, and another 130,000 marched in Australia in November 2007.

Everyone knows now that something must be done. But nothing remotely adequate is being done. We are being drowned in greenwash. Everything has changed and everything stays the same.

Part Four explains how this contradiction has come to be and how we can think of breaking through it. This chapter is about climate politics in the 1990s and the negotiation of the Kyoto Protocol. This is an inspiring story, because the climate scientists have managed to alert the world. But it is also a story that shows the limits of trying to stop global warming by lobbying governments.

Climate Scientists

Most people assume that the climate movement began with environmentalists. In fact, scientists were the driving force for many years.

163

In the late 1980s several hundred scientists around the world, in many different disciplines, were working on topics connected with climate change. They were beginning to get a clear understanding of the threat that faced the planet. These specialists came together in a series of international scientific conferences. In the corridors and the bars they talked to each other about what to do.

They set out to alert the people of the world to global warming, and to persuade governments and the UN to act. They first organised themselves, and then reached out to educate and involve scientists of every kind. The environmental NGOs, like Greenpeace and Friends of the Earth, then helped get the message out and tell the world what the scientists were saying.

What the climate scientists were doing was quite new. There had been political movements of radical or concerned scientists before. But there had never been an organised movement of all scientists with a major political agenda. The climate scientists did not do this because they were radical people. They were not. But they were the first people who could see clearly what would happen if the world did not act.

In 1989 the climate scientists also took a fateful decision. They decided to organise in alliance with sympathetic governments and politicians. They did this to change the minds of the powerful. This alliance had two consequences. The scientists succeeded in alerting the whole world to what was happening. But the alliance with the governments also meant that the scientists could not achieve the solutions they knew were necessary.

Let's look at the story of that alliance.

In 1989 the scientists founded the Intergovernmental Panel on Climate Change (IPCC), under the auspices of the United Nations. The IPCC produced major reports every few years—in 1990, 1995, 2001 and 2007. Every report consisted of three long volumes—one on the basic science, one on the likely effects of climate change and one on what could be done to slow it down. The scientists were making an argument—this is what is happening, this is what will happen and this is what we can do about it.[1]

Each chapter of each volume was written by a panel of scientists from all over the world, with government representatives sitting in on all the drafting meetings. The climate scientists educated these civil servants. In the process, they recruited people inside each government machine who became champions of action on climate change. Many of these champions were scientists, or had chosen to work for the environment. In other words, they were the people

Climate Politics

most likely to be won over. By going through these people, the scientists were also trying to compel governments to act. When the reports came out, they had been agreed by people from the ministries in all the major governments of the world.

But there was another side to the IPCC. The American government delegates brought Big Oil and Big Coal into the room.

The carbon corporations

In the 1990s the climate scientists and their champions were clear that action over global warming would have to cut emissions globally. This was equally clear to oil, gas, coal and car companies. From 1990 on the companies organised to stop action over climate change.

The "Carbon Club"[2] of coal, oil and gas companies had a three-pronged strategy. The first prong was to convince the public and politicians that global warming was not happening. The second was to convince Bill Clinton's administration that cuts in CO_2 emissions were against the interests of American industry. The third was to use the power of the American government to prevent a global agreement to limit emissions.

Peabody Energy, the largest coal company in the US, took the lead in the Carbon Club through most of the 1990s. Peabody was careful to keep it that way. But coal is the largest source of energy for electricity in the US, and coal emits more CO_2 than oil or gas.

Big Oil followed Peabody's lead. The leader here was Exxon Mobil, the largest oil company in the world. Until the late 1990s the rest of Big Oil, including Shell and BP in Europe, were also an active part of the Carbon Club.

By the late 1990s the balance shifted. Shell and BP, under political pressure in Europe, retreated from open identification with the Carbon Club. But Exxon Mobil, with close connections to the new Bush administration, took over the leadership from Peabody Coal.

The Carbon Club concentrated their campaign in the US. Big Coal and Oil had more power there than in other major country. The US also mattered more than any other country in global climate politics, with the US responsible for almost a quarter of global emissions. Just as important, the US government was in the habit of doing what business wanted, and it was the dominant military and economic power in the world. The US government was the leading voice in the UN, the World Bank, the WTO and global negotiations of every sort. And because of American power, American ideas dominated the world. All that made the US the critical battleground.

For the carbon corporations the first line of defence in the US was denial. The scientists were trying to convince the world that global warming was real and dangerous. If that was true, it followed that something had to be done. The carbon corporations set out to convince the American public that there was no scientific consensus on global warming.[3]

The carbon corporations paid individual scientists quite large sums to act as spokesmen and "climate sceptics" or "deniers". The money was channelled through foundations and think tanks, so the scientists looked independent. On a scientific level, this strategy was an almost total failure. Few scientists were recruited, and very few were climate specialists. Their work was shoddy, since they were defending shoddy positions. They did not convince other scientists. They could not get their papers published in reputable scientific journals.

But at the level of public opinion, the strategy worked well for years. The American media—television, magazines and newspapers—consistently ran stories that treated climate change as an issue with two equal and conflicting points of view. They would quote a mainstream scientist and then one of the deniers. That made it seem as if the science of global warming was muddled, unclear and inconclusive.

Americans grew confused. Answering a *Newsweek* poll in 1991, 35 percent of Americans said that global warming was a "very serious problem". By 1996 only 22 percent told *Newsweek* it was a very serious problem. The intervening five years had seen the scientific evidence grow more convincing and more worrying. But it had also seen extensive climate denial in the media.[4]

The stated reason for the media coverage was "balance". But balance was an inappropriate model. There are people, for instance, who deny that American astronauts ever reached the moon. The newspapers do not balance their opinion with NASA scientists. Nor do they now cover the link between smoking and cancer that way.

The American media are controlled by large corporations, each of which owns many newspapers, or magazines, or television stations, or radio stations, and often all of those. One seventh of media advertising is paid for by the car industry.

Some have blamed journalists for the state of media coverage. But the pattern is too uniform to be just down to lazy or overworked journalists copying press releases. This was the sort of story their bosses wanted.

Climate Politics

Denial was the first line of defence. The second was persuading the White House and Congress that even if global warming was real, doing anything about it would hurt American business. In the 1990s this was not hard. Campaign contributions helped, and so did the influence of Big Auto with many Democrats. But the most important thing was that every American administration had always supported the international interests of American corporations. If Big Oil, Coal and the car producers wanted the White House to support them in climate negotiations, that would happen.

For a moment in 1992 it looked like President George Bush, the father of George W, was prepared to consider doing something about global warming. Then Sununu, his chief of staff and a former electricity executive in New Hampshire, stepped in and Bush agreed to the corporate line.

Compromise in the IPCC

However, the dominance of the carbon corporations in the US was not the end of the story. They had nothing like the same impact in Europe and the poor countries.

This is because ruling class people—the corporate executives and politicians who run the world—are not stupid. Outside of the fossil fuel and car corporations, they were beginning to listen to the scientists. But the American government brought Big Coal and Big Oil into the IPCC. Jeremy Leggett, the lead climate campaigner for Greenpeace, was appalled when he arrived at the IPCC meetings in 1990. He found men from Peabody and Exxon sitting in the meetings alongside the American government delegates. The coal and oil men had their laptops in front of them, and they edited every line of text.[5]

Right through the 1990s there was a four-way struggle in the IPCC. On one side were the scientists, on the other coal and oil. In the middle were the governments of Europe, which wanted to do something, but not too much. On the sidelines, with very little power, were an alliance of the environmental NGOs, the African countries that faced famine and drought, and the small island countries that would go under the waves. They backed the scientists wherever they could.

Government representatives only sat on the bodies that drew up the summaries of each IPCC report. The summaries were what they really cared about, because that was the part journalists and politicians read. So, in theory, the scientists could put the unvarnished

truth in the main text. In practice, however, the scientists were looking over their shoulders with every word they wrote.

The scientists were in the strongest position when writing Volume
One of each report, on the basic science. But the governments won
the battle over Part Three of the reports—what is to be done. Here
the reports have been done by environmental economists. These are
liberal people compared to other economists, but it is a deeply conservative profession.

Environmental economists work with the idea of a perfect market distorted by "external" factors. Their aim is to adjust the market
so it will solve environmental problems. The environmental economists want to persuade governments that they will save money in
the long run by spending money now on climate measures. So they
work out how much the effects of global climate change will cost.
Then they work out how much governments will have to spend to
avoid these disasters.

One problem with these calculations is that the numbers are
unreliable. No one knows what the effects of climate change or the
alternative solutions will cost. Another problem is that the numbers
the economists use count only the costs in dollars to property and
companies, not the damage to human beings.

Moreover, the economists' solutions stay within the limits of the
market. They talk about ways to make cars more efficient, and ways
to promote public transport, but not about banning cars. They talk
about making cement more efficiently, but not about controlling
how much cement is made. They talk about "benchmarking" steel
plants to see what technologies are most efficient, but not about
closing inefficient plants and opening new ones. The economists do
mention regulation from time to time, but only in passing. What
they can't do within their framework is talk about the kind of radical changes I have outlined.

Another problem is that the economists assume that what the
governments care about is not spending money. This rules out
expensive solutions. More important, the economists cannot see that
neoliberal politicians and corporations do not look at climate policy primarily in terms of how much it costs. They are willing to
spend very large sums of money on some things, such as the Iraq
War. What they hate about spending money on climate measures is
just not the expense. It's the precedent of using government money
to meet human needs, and interfering with the power of the market.

For all these reasons, the approach of the environmental economists has failed to persuade governments to act on the necessary

Climate Politics

scale. But what it has done is to limit the kind of solutions that could even be thought about. Unfortunately, their economics have not just affected the reports of the IPCC. They have structured the whole way everyone talks about how to halt global warming. The compromise, then, was that the scientists got to warn the world about the dangers of global warming. The governments and the economists got to control the answers to the question of what to do.

The weakness at Kyoto

As the 1990s went on, the scientists, and their champions in the ministries, pushed for concerted global government action. Three central things were clear to those who wanted action. One was that the only action that would make a difference would mean somehow using less oil, gas and coal. The second was that action would have to be global and involve all the major countries of the world. The third was that this would mean agreements between governments for each country to cut its total emissions.

In 1992 the United Nations convened a global conference in Rio de Janeiro on the environment for scientists, campaigners and governments. The Rio conference agreed on international negotiations leading to a global treaty to limit CO_2 emissions. After five years of negotiations an international conference in Kyoto, Japan, finally produced a preliminary agreement, the "Kyoto Protocol". But negotiations on the key provisions continued for another four years, until a conference in the Hague agreed a final text in 2001.

That final text was fatally flawed, in ways I will describe later. On the surface, the Kyoto agreement was weak because the carbon corporations and the US government were so strong. They were only strong, however, because the people fighting them were weak.

The results of negotiations are almost never determined by what goes on at the meeting. What matters is what power each party brings to the table. At Kyoto and the Hague the international negotiations were gatherings of the elite. There were no mass movements back home to force governments to act, and no representatives of the social movements at the negotiations. The environmental NGOs were present, but their strategy was to persuade governments to act. So they relied on the governments of Europe and the poor countries to push the case forward, and were disappointed. Much worse, the social movements—the unions, the political parties, the campaigns and the faith groups—were not even present outside.

Both the environmentalists and the social movements have now begun to change. 2007 is not 2001. There is hope now. But to understand the situation we're in, we need to look squarely at the weaknesses of our own side. I will start with the environmentalists.

The Environmental NGOs

On a global scale, the largest environmental NGOs (Non-Governmental Organisations) are Greenpeace, Friends of the Earth, and WWF (formerly the World Wildlife Fund). In the US the Sierra Club, Environmental Defence and Energy Action are massive organisations, and there are thousands of other environmental NGOs around the world. The strategy of the environmental NGOs in the 1990s went in parallel with the strategy of the scientists. The NGOs publicised to the world what the scientists were saying. Together, they have told us. Without this achievement there would be no possibility of averting catastrophe. However, the NGOs also concentrated, like the scientists, on bringing about change by lobbying governments. During the 1990s many NGO campaigners discovered at first hand the weakness of such a strategy. This is why the NGOs are now moving towards more activist strategies.

To understand the NGOs, you have to start with a contradiction. The campaigners care passionately, and have dedicated their lives to saving the environment. But they are constantly subject to pressures that force them to accommodate within the limits set by the rich and powerful. The passion is obvious and does not need explaining. The pressures on them are less obvious and more confusing to the NGO campaigners themselves.

The model for almost all environmental NGOs is the traditional North American lobbying campaign. Each NGO is basically a group of full time professional campaigners paid from money raised by the NGO. The fulltimers in turn produce reports, get the message into the media, lobby the government and politicians, and sometimes take corporations or the government to court.

Like most North American NGOs, the environmental campaigns survive on donations, mostly from affluent professionals (though sometimes also from governments). The largest donations tend to come from the richest donors. This financial base exerts a constant pull on the campaigners. Every policy position and every sentence in leaflets, publicity and press statements has to be read carefully for its impact on the donors.

Climate Politics

The environmental campaigners believe that their donors are well to the right of them. As we saw, for instance, one reason environmental NGOs were so slow to campaign against SUVs is that they knew many of their donors drove them. In the 1970s Greenpeace in North America found to their cost that they raised a lot of money when they campaigned against French nuclear tests and much less when they challenged American nuclear tests or Canadian sealing.[6]

The divide between donors and campaigners is not that clear cut. Most NGO staffers come from the same background as the donors. In the US they tend to be graduates of elite universities, often in the liberal arts, sometimes in environmental sciences, rarely in engineering. In Britain they usually went to "public" (that is, private) schools. In both countries they share many of the bedrock assumptions of the class they come from. This does not mean they are right wing. Indeed, they have rejected the greed they grew up with and want a better world with different values. But the politics they hold are a compromise between corporate values and an alternative world. For most of them, too, the environment appeals on a deep emotional level as a place of peace. It's the place where humanity lived before industry. It's a place where corporations, factories and the urban working place are absent. They feel that nature is good, and humanity greedy and imprisoned by growth.

To simplify a very complicated reality inside the heads of a large variety of people, these professionals reject the world the corporations made. But they do not usually see the world as a conflict between workers and corporations. Instead what they want is an alliance of all decent people, like themselves, across the normal divides of class and politics. This human desire to build alliances of decent people fits with their donors, who feel the same. It is also reinforced by the project of lobbying. Green NGOs are run by realists. But there is realism and realism—it all depends on what you think reality is. Green NGOs look to the people at the top to change things. That's not because they like the people at the top, but because those people have great power.

It's also because the green NGOs have faith in democracy. The environmental movement began in the early 1970s with a democratic model of American and Western European society. Their strategy was to alert the public to serious problems, believing that once the public understood, citizens would use the power of the ballot box. Politicians were centrally concerned with winning elections, they reasoned, so convincing the public would change government policy. This is an understandable view of the world, the one taught in schools. The difficulty is that "democracy" does not work like this. But at first the strategy seemed effective.

In the United States the environmental movement exploded with the Earth Day teach-ins and demonstrations in 1970. It was the last of the big movements of the 1960s. The Nixon administration, already badly battered, did not feel capable of standing against yet another movement. Environmental controls were tightened, and the air grew cleaner. Then Greenpeace, with courage and brilliant showmanship, were able to save the surviving species of whales. In the 1980s, after Three Mile Island and Chernobyl, new construction of nuclear power stations was effectively stopped. In other areas results were more mixed. Many toxic waste dumps and incinerators were resisted, but some went ahead. Many forests and wetlands were saved, but many were lost.

But in the 1980s the tide of global politics began to flow against the environmental NGOs. The Reagan administration was openly hostile and governments around the world were doing little. Because environmental NGOs were committed to lobbying, this more general political shift pulled them to the right. In Britain, for instance, it meant that all serious NGOs put a lot of effort into influencing the Conservative environment minister of the day, and later into a desperate hope that New Labour would be better. In the US it meant a reliance on the judges in the 1980s and 1990s, and Clinton and Gore in the 1990s.

Environmentalists usually defended this swing to the right by saying they had to argue in ways that would be taken seriously. They could not take this or that position, use this or that language, tell this or that brutal truth, because if they did they would not be taken seriously. It was usually put like this, in the passive—as not being "taken seriously"—rather than saying who it was who would not take us seriously. This was because quite radical positions on the environment could command majority support, but the people at the top would not deal with you if you went too far.

By the early 1990s many environmental campaigners were wrestling with what they saw as the failure of the democratic model. On issue after issue, they had convinced the world. But they could not move the politicians. Later in the decade even the politicians they put their hopes in, like Clinton and Blair, would betray them. This was a much more general problem than just for environmentalists. By the late 1990s it was called the "democratic deficit". But for environmentalists it was especially challenging, because they were trying to reason with the people at the top.

In this situation environmental campaigners faced difficult choices. They could bend to fit the realities of power, and talk the language of corporate business. For many this was painful. Or they could organise a mass movement outside the accepted political parties, in the way that civil rights groups and trade unions had once done. But the NGOs couldn't do that—their established habits and traditions ruled it out. And in the early 1990s mass campaigns seemed utterly unrealistic to the NGO fulltimers. So by and large they bent to power. This meant that when it came down to the negotiations over Kyoto, the environmental NGOs had a place in the meetings, but could bring little countervailing power to the table.

The scientists had a power base in their expertise. They had an alliance with champions of climate action in every ministry. And the ruling class in many countries was genuinely split over what to do. The carbon corporations had real power, especially in the US and the Middle East. Each of these groups could bring real power to bear if, from their point of view, the negotiations went wrong. The environmental NGOs could threaten to criticise, but they could not threaten to leave. Nor did they have any sanctions against government and corporate power back home. They had not built a movement to do that, and it was not part of how they worked.

Global warming also presented environmental NGOs with a new problem. They were used to campaigning for things that mattered, but on a much smaller scale than global warming. It was possible to save the whales without challenging the world economy. Stopping one incinerator pitted them against a single company. Getting a clean air act passed pitted the NGOs against whole industries. But the industries could live with a clean air act.

Climate change was a much greater threat and required a structural response that would change the world economy. Lobbying would not be enough. To make a difference, the environmental NGOs would have to find new ways to act. By 2001 almost all NGO campaigners knew this. The failure of Kyoto also led many to rethink their strategy. At the climate talks in the Hague where the final agreement over Kyoto was concluded in 2001, Friends of the Earth International had 5,000 demonstrators protesting outside. NGO activists began looking to the new social movements for inspiration.

The absence of the social movements

When I speak of social movements here, I mean not only the campaigns and the unions, but also the far left and people like the left of the Labour Party in Britain, the Refounded Communist Party in Italy, the CPI in India and the more liberal Democrats in the US. These are the sort of people who are the backbone of Stop the War in Britain, United for Peace and Justice in the US, and the World Social Forum.

Unlike the environmentalists, most of these people do know how to organise big demonstrations and a mass grassroots campaign. Until very recently they have done almost nothing about climate change. This has been a very important absence. The oil companies and the American government were partly able to dominate the Kyoto negotiations because the environmentalists were weak. More important, the unions, social campaigns, opposition political parties, churches and mosques were not involved at all. This was a far worse dereliction of duty. But it seemed so normal that it was hardly commented on.

There are several reasons for the silence of the social movements. The first is a long history of opposition between the unions and environmentalists, particularly in North America.[7] In one controversy after another in the US the environmentalists have been on one side, while the unions were on other, alongside the companies, defending their jobs. The environmentalists have felt that ordinary people are the problem, and the union members have felt the greens are arrogant rich kids.

None of this, however, lets the unions and the social movements off the hook. Their deep weakness on the environment was partly a reaction to green hostility. But it was also part of a more general failure of nerve. In the US, for instance, the neoliberal attacks meant that by 1990 the unions were constantly on the defensive. Individual union activists still believed in the values of solidarity and sharing, and when they got drunk they told each other so. But it had become publicly embarrassing to speak of that old time religion.[8]

Union members, and union leaders much more strongly, felt that the only way to survive in the hostile new atmosphere was to function as defensive interest groups. So timber workers defended clear cutting, car workers defended SUVs and miners defended coal.

Working people still love the environment and are well aware they get the worst of any poison and pollution going. But when directly threatened, they defend their jobs in alliance with the cor-

Climate Politics

porations. This ambivalence produces abstention and guilt, not action. It is this that allowed Bill Clinton and then George Bush to say they would not take action on climate change because it would threaten American jobs. And it was this that allowed the Senate to vote 95 to 0 against a climate treaty in 1997.

In Europe and Latin America there is another historical reason that many union and social activists shy away from environmental activism. During the 20th century communists led many of the union and social movements in these countries. Back in the 19th century, Karl Marx wrote about green issues in ways that would be familiar to modern environmentalists. But 20th century communists looked to the example of the Stalin's Soviet Union and Maoist China. Both had appalling environmental records, because the dictatorships used and abused the land in the same way they used and abused their workers and peasants. This record continues, incidentally, in the market authoritarian democracy of Russia and the free market dictatorship of China.

But there is another factor as well. Many activists accept the way right wing environmentalists like James Lovelock frame the argument.[9] Like Lovelock, they also think the choice is between industry and nature, between jobs and biodiversity, between decent living standards and the atmosphere. For union and social activists, this choice, conceived in that way, is paralysing.

But the social and union activists have now begun to get on board the climate train. Some of them have played an important part in organising climate demonstrations in many countries. In early 2008 the Campaign against Climate change in Britain organised a conference for trade unionists, and 300 activists came. This is happening because the project begun by the climate scientists in 1989, and popularised by the environmental NGOs, has done its work. Everyone knows now that something must be done.

The flaws in Kyoto

Let's look now at the agreement that came out of negotiations at Kyoto and the Hague. The scientists and NGOs had set the stage. The scientists had told everyone what was happening. They had convinced many politicians and corporate leaders that they had to do *something*. But because the scientists and NGOs were trying to convince the people at the top, they had very little control over *what* that something was.

There were basically four positions in the ongoing global negotiations. One was the position of the American government—they did not want a treaty. The carbon corporations had won the argument with American politicians. They were committed to only signing a treaty that commited poor countries to reducing their emissions as well. As the treaty on offer was not going to commit the poor countries, all American politicians, Democrat and Republican, stood united to do nothing. That did not stop the American delegation from trying to dominate the negotiations in Kyoto and the Hague. They used that domination to make the treaty as weak as possible, and then refused to sign it. This meant that the emissions from oil could continue spewing out, not only in the US, but round the world.

The second position was held by the European governments. They wanted a stronger agreement, but one that did not ask them to do too much.

The governments of most of the poor countries held the third position. They wanted the rich countries to take action on climate change. Their governments were well aware that their people would feed the main effects of global warming. They refused, however, any treaty that required them to take action. Economic growth and escape from poverty came first.

The fourth position was held by those small island nations, like Tuvalu and the Maldives, whose lands would soon go under the water. They were desperate for action and had no power.

The NGOs—Greenpeace, Friends of the Earth and the others—were allowed into the public sessions. They issued bulletins to the world press, detailing the dangers of the compromises that were being made. Wherever possible, they tried to work through the island nations, the poor countries and the European governments to block the proposals of the United States and the carbon corporations. But because they were looking to some governments to stop others, they were up against the most important fact of global politics in the years from 1992 to 2001. The United States government and corporations dominated the world in those years and so they dominated the global climate negotiations. The key meetings took place late at night, by American invitation. The chairs who ran those crucial meetings were chosen by the American delegation, while the delegates from the poor countries, the island nations and the NGOs found themselves locked out of the room.

The central aims of the US negotiators were to reduce the cuts for

American corporations and to protect the coal and oil industries. So in negotiations they insisted on three changes to the proposed treaty. They wanted the total amount of cuts reduced. They wanted any method of legal enforcement of the treaty removed. And they wanted "carbon trading".

At first the delegations from Europe and the developing countries resisted all three because they would reduce the total amount of emissions cut. Then they capitulated, in order to get the Americans to sign up. After Kyoto the Americans refused to sign, but the damage was done.

Kyoto works like this. In the first stage, to run until 2012, only the richer countries agreed to cut their emissions. These countries are the USA, Canada, all of Europe, Japan, Australia and New Zealand. All of them except the USA and Australia eventually signed the agreement. The poorer countries do not have to cut their emissions in this first stage.

Negotiations have now begun for a second stage, which will begin in 2013. By 2012 all the richer countries that signed should have to cut their emissions by a certain percentage. The European governments and the poor countries wanted cuts in CO_2 emissions of 10 percent or more. The actual numbers vary from country to country—some negotiated hard. But the final treaty provided for average cuts by the richer countries of 5.2 percent by 2012. This means that the average CO_2 emissions for each rich country in an average year between 2008 and 2012 have to be 5.2 percent less than the emissions for that country in 1990.

In reality, most countries have to make greater cuts. By the time the first draft of Kyoto was agreed in 1997, most countries' emissions were already a good deal higher than in 1990.

The only exceptions were the countries of the former Soviet Union. There the base starting date—1990—was just when the "market transition" from Communism was beginning in Eastern Europe and Russia. That transition destroyed much of the industry in those countries and slashed living standards. By 1997 the Eastern European countries had already made very deep cuts in carbon dioxide emissions, inadvertently.

The governments negotiating Kyoto knew very well that Eastern European industry and emissions had collapsed. This was the background to "carbon trading" in Kyoto, one of the treaty's greatest flaws. Carbon trading allows one country which has not made the required emissions cuts to pay another country for part of their emissions quota. When Kyoto was negotiated, everyone knew that this meant that Western Europe and Japan would carbon trade and pay Eastern Europe for the privilege.

The result of this carbon trading, inevitably, is to increase the total of emissions. Let's imagine, for example, that Britain does not make the cuts in emissions it agreed to. So Britain has to buy permits for, let's say, 100 million tons of CO_2. But Ukraine actually saw its emissions fall by more than 100 million tons after 1990, because Ukrainian industry was in trouble. So Britain pays Ukraine for 100 million permits. Ukraine has more money. But Britain is still emitting 100 million more tons than it promised, and Ukraine is emitting exactly as much as it would have otherwise. The possibility of carbon trading has increased total emissions by 100 million tons. This is not some imperfection of the market. It is what the people negotiating Kyoto knew would happen. The American delegation insisted on carbon trading because they wanted loopholes.

The American delegation also insisted on taking out any enforcement mechanisms in the treaty. There would be no penalties, and no courts. This gutted the treaty. In 2007 the Canadian government announced they would not make their targets before 2025. Most countries in Western Europe were well over their targets in 2007, and their emissions were rising. They won't meet their Kyoto targets either because they don't have to.

The final thing the American delegation insisted on was that the treaty would not come into force until countries responsible for 55 percent of the total global emissions in 1990 agreed to implement it. In 1990 the United States was responsible for almost a quarter of global emissions. The poor countries which were not implementing the agreement accounted for about a fifth. So this clause meant the treaty would not come into effect if only the USA, Australia and Russia refused to implement the treaty. The American government was confident that Australia, the world's biggest coal exporter and a close American ally, would not sign. They were also reasonably sure that Russia, a major gas and oil producer, would not sign. So the 55 percent clause effectively gave the US a veto on Kyoto ever happening. That's why that clause was in there.

When all the negotiations were finished, the US government and the oil and coal corporations had a weak text that couldn't work without them. The treaty was finally agreed at a meeting at the Hague in 2001, and the US, Russia and Australia promptly refused to implement it. It looked like Kyoto was dead.

Then late in 2004 the Russian government reversed itself and signed to everyone's surprise. In February 2005 Kyoto came into effect. One reason the Russian government changed its position was carbon trading—the Russian economy was still producing way below 1990 levels and Russian companies stood to make money. The other reason was that global politics had changed. Putin could defy Bush in 2004, as he could not in 2001. All the same, when Kyoto did come into force, the clauses the American government and the oil corporations had fought for meant that it was weak, unenforceable and full of market loopholes.

But there is a case to be made for Kyoto. First, at least the treaty was there. The governments of the world had done something quite new. If the pressure could be kept up, the next treaty after 2012 could cut emissions far more deeply. Without Kyoto there would be nothing. Second, the 5 percent cuts in emissions were not nearly as weak as they sound, because these were cuts on the level of emissions in each country *in 1990*. Back then global emissions were only 22 billion tons of CO_2 a year. Now they are 28 billion tons. It would take a cut of 25 percent now to get down to 5 percent below 1990 levels. Third, the existence of the treaty has meant that many governments in the richer countries have made more effort to hold down emissions than they would otherwise have done. Most of the developed countries are not meeting their targets under the treaty. But the real growth in emissions has been in those countries that did not sign, like the United States, or did not have to sign, like India and China. Finally, there will be no way of cutting CO_2 emissions in the future without a global agreement. Kyoto was a start.

What the scientists and environmentalists had done, starting from where they did and with the weapons they had, was a gift to humanity and the planet. It was also not nearly enough. By 2007 many scientists and environmental activists were beginning to think they would have go a lot further, and find a way to force the governments and corporations to act.

They were able to start thinking like that because the new social movements of the 21st century had changed global politics.

CHAPTER 15

Climate Politics after 2001

THE new social movements exploded into the consciousness of the world with the protests at the World Trade Organisation summit in Seattle in November 1999. To understand these movements, we have to start with the feelings they came from.

By 1999 people all over the world had lived through 20 years of neoliberalism—of privatisation, union busting, welfare cuts, bullying and stress. Almost everywhere in the world almost every job was harder than it had been a generation before. People had lived through big defeats for mass movements too, and little defeats in their own lives that hurt like hell. Neoliberalism and globalisation transformed the details of people's lives. They made people anxious, tore at the edges of marriages, and ground down the idea that kindness to others was the central value of a human life.

All those years exploded in the new social movements. There were two striking things about the new movements. One was that most of the marchers and activists were young. The other was that their opinions, and their hatreds, were no different from those of their parents. They had lived through the last 20 years together. Mothers and fathers had passed on to the children their understanding of what had happened, their helplessness and their bitterness. This was payback time.

But one thing changed between the generations—the idea that you can't fight back and win.

The breakthrough was the demonstrations at the World Trade Organisation summit in Seattle.[10] The first important thing about Seattle is that it was a demonstration of workers against neoliberalism. The majority of the 60,000 protesters were union members brought by their unions. They were demonstrating because the WTO was meeting to break down world trade barriers as part of globalisation. For 20 years American unions had been told there was no point in fighting back, as the company would just move the factory. So they were demonstrating against the world market.

The second important thing is that they won. A bit under half the demonstrators were environmentalists, anti-poverty campaigners and young activists. They surrounded the convention centre where the WTO delegates were meeting and tried to have a non-violent sit-down blockade. The Seattle police attacked with clubs and gas. The union members were marching separately, but some of them broke away and through the police to join the young people. The streets filled with gas. Angry protesters began throwing rocks through the windows of the chain stores, Starbucks and McDonald's. The delegates, frightened, did not go through the demonstrators.

The WTO—the headquarters of globalisation—was closed for a day. When it reopened, the American government had lost control of the delegates. The people from the poor countries had been willing to do what President Clinton and American business told them to do. Now they thought Clinton can't even control his own people, so why do we have to do what he says? At the end of the summit the delegates left, having refused to make any new world trade agreement. The protesters had won.

That victory resonated around the world. There are many claims for the beginning of the new global movement. It can be dated to the rising by the Zapatistas in southern Mexico in 1994 or the public sector general strike in France in 1995. But Seattle was the moment when the world understood that politics had changed. That was partly because it was in the United States. After all, America is the superpower, the dominant political and cultural force in the world. When things happen in America, people notice.

But it was also because of what the protesters were targeting. They were thinking big. They were against the system. And they tried for a moment to close it down. They were against the world market that everyone else had grown to hate.

This was not because everyone in the world suddenly had a Marxist analysis of global capitalism. It was because the politicians and the corporations and the media told everyone, over and over again, that they had to do what the world market said.

In the 18 months after Seattle activists scrambled to organise protest marches outside other summits. There were substantial demonstrations in Melbourne, Port Moresby, Seoul, Durban, Nice, Gothenburg, Barcelona, Prague, Zurich, Washington and Quebec City. The largest was at the G8 summit of world leaders in Genoa, Italy, in June 2001. And the movements in all their variety came together for a series of global conferences, notably the World Social Forums in Brazil, India and Kenya and the European Social Forums in Italy, France, Britain and Greece.

The street marches and the forums had two central slogans. "Our World is Not for Sale" defied neoliberalism. "Another World is Possible" marked another watershed. Since the 1980s the idea of an alternative to capitalism had been dead. Now we were shouting in the streets that we could change everything, the whole world.

For 30 years opposition politics had been dominated by identity politics. Everyone had their own struggle—women, gays, lesbians, immigrants, African-Americans, native peoples, Muslims and so on. The campaigns were separate too—environmentalists, unions, the peace movement, and so on. But the new movement insisted that what mattered was not what divided us but what united us.

Our causes were myriad. But all the forces we were fighting against—neoliberalism, racism, sexism, war and the destruction of the environment—had their roots in the same global capitalist system. As we met at protests and in forums, environmentalists trying to stop GM crops and tenants campaigners fighting for public housing explained to each other how it was all one struggle with one system.

These changes in the shared political common sense of radicals around the world were crucial for fighting climate change in several ways. For one thing, it was possible to imagine a global struggle. Second, an alliance between the environmental movements and the social movements was now possible. Third, if our world is not for sale, governments can intervene to save the planet. And if another world is possible, then it is possible to stop climate change.

War

More crucial developments for the fight against climate change grew from George Bush's War on Terror, and the global resistance to war.

When Bush came into office in 2001, his government represented the oil companies. Vice-President Dick Cheney, the former head of the oil field services company Halliburton, was the most influential figure in the administration. From the first day in office Cheney had two priorities. One was a comprehensive energy review to shift American government policy. This produced a plan for hundreds of new coal fired power stations, new nuclear power stations, and new tax breaks and drilling rights for the oil companies. Along with this went a complete denial of global warming. Clinton had not campaigned to support Kyoto. Bush withdrew American support and made himself the global leader of opposition to action on the climate.

Cheney's other long term strategy was to regain control of the oil countries in the Middle East. That meant breaking the governments of Iraq and Iran. When Bush was first elected, however, it was politically impossible to attack Iraq or Iran.

Then came 9/11. From Bush and Cheney's point of view, 9/11 was a threat, but even more an opportunity. It was a threat because the spectacle of the Pentagon in flames was a direct challenge to American power. The US was supposed to be the greatest power on earth. That power rested in part on fear. That fear would have to be restored. Many Middle Easterners would have to die and America would have to win a war. Afghanistan, devastated by 23 years of invasion and civil war, was the easy target.

At first that looked like an easy American victory. Intoxicated by success, Bush, Cheney and the defence secretary decided it was time to "do" Iraq.

There was much at stake. The United States was still the world's leading economic power, but no longer dominant. The European Union had as big an economy, Japan remained a major economic competitior and China was a rising economic and military power. The US, however, was the world's dominant high-tech military power. If they could win a bombing war in Iraq, that would make them the dominant world power.

Then there was the oil. Iraq was said to have the second largest oil reserves in the world. The plan was to privatise those reserves after the invasion, and offer them to American oil companies. After that, once Iraq was subdued, they could "do" Iran. It would be, in the neoconservative phrase, the "American Century".

There were three flies in that ointment. All three have an important bearing on global warming.

First, the other major global powers did not back the US invasion of Iraq. This was because it was obvious to them that it would mean American dominance of the world, and they did not want that. They had more or less approved of the invasion of Afghanistan. But when it came to the invasion of Iraq, France, Germany, Russia and China were all opposed. There are constant noises that Germany and France will somehow return to the fold, but they have not yet done so. So the Iraq invasion marked the moment when US hegemony began to unravel.

The second snag for Bush and Cheney was the global resistance to the war. This was weakest in the US, for entirely understandable reasons. If something on the scale of 9/11 had happened in London or Paris in 2001, British or French people would also have rallied round a right wing government intent on revenge. That would not necessarily be the case now, in very different political circumstances.

But in 2001 the American social movements felt the ground cut from underneath their feet, and it would be four years before the peace movement was really strong in the US.

In the rest of the world, the opposition to the war began with the activists in the new social movements. The anti-war movement was particularly strong in Britain, Italy and Spain, because their governments were going to send troops. As war loomed, the European Social Forum met in Florence. The leading figures were the new leaders produced by the Genoa demonstrations. They called a demonstration against war at the end of the ESF, and a million people came from all over Italy.

Before Florence, at another organising meeting for the ESF, British and Italian activists persuaded other movements to call global demonstrations in every country against the war on 15 February 2003, just before the invasion.[11] Two million marched in London, two million in Rome and three million altogether in many cities across Spain. Simultaneous demonstrations in many other countries involved a total of something like 20 million marchers. This was the largest global demonstration ever, about anything. It established, and opinion polls confirmed, that the majority were against the war in every country but Israel and the United States. By 2005 opinion had turned in the US as well.

These marches had a lasting effect. They meant that people in the Middle East, Pakistan and the United States knew that the world was opposed to the American invasions. That did a great deal to encourage the resistance in Iraq and Afghanistan. It also gave heart to the American peace movement. The marches made it politically difficult for the US to reinforce its troops in large numbers. When the going got tough, the Pentagon went for a surge, but in small numbers and for a limited time.

The peace marches had three important effects on climate politics. First, environmental activists were involved. In Britain, for instance, the Green Party was part of the Stop the War Coalition. Greenpeace and Friends of the Earth, more careful of their NGO status, were not. But on the day of the big march in London the Greenpeace full-timers and activists met outside their offices and marched off to join the demonstrations, banners flying. I also have no doubt that every Friends of the Earth fulltimer was on that march. More important, so were tens of thousands of environmentalists.

Globally, hundreds of thousand of individual environmentalists now felt themselves to be protesters. The peace movement also started them thinking about a global climate movement.

Climate Politics

The peace movement and the growing defeat in Iraq have also had important consequences for general politics, and therefore climate politics, in America. It weakens the oil companies in American politics. It has discredited Bush, and the whole neoconservative project, in front of Americans. That means the climate deniers have lost confidence—Hurricane Katrina had a massive impact here too—and Al Gore can be heard. In fact, the whole project of neoliberalism is under attack. All the Democratic presidential candidates are talking left, and none of them are talking neoliberalism the way Bill Clinton did in the 1990s. 2006 also saw massive demonstrations for immigrant rights. A million marched in Los Angeles. More startling, 400,000 marched in Dallas, the most right wing city in Texas. After the Democrats took Congress, there were three bills in Congress to do something about global warming, a bill to reduce gasoline mileage limits in cars, and efforts in cities and state legislatures across the country to start limiting CO_2 emissions.

The looming American defeat also has major implications in global climate politics. American power has not disappeared. But Western Europe, China and Russia are distancing themselves from Washington, and establishing alternative centres of power. When Bush announced in 2007 that the US would lead other countries in an international response to global warming, it was a joke. At the G8 summit in Germany in June 2007 Bush was isolated on climate change. It now looks possible that the other world powers will push ahead with negotiations with some very flawed successor to Kyoto.

Nothing is certain in global politics. But if American defeat does finally come it will have a shattering effect. Partly this is because most people don't imagine the US could be defeated. But it's also because the whole project of neoliberalism and globalisation has been presented as an American led package, the "Washington consensus". With an American defeat in Iraq, people all over the world will think—well, if the Iraqis could do that to the greatest power in the world, maybe we can stand up to neoliberalism.

Beyond that the US would lose control not just of Iraqi oil, but of all Middle Eastern oil. The dictatorship of the Saudi royal family depends on American support, as does the Mubarak dictatorship in Egypt. If the Americans leave Iraq, all the dictators will tremble. The will try to distance themselves from American power and are likely to be overthrown by popular uprisings.

But from Washington's point of view, it's worse than that. An American defeat in Iraq would also end US domination of the world economy. It could even spell the end of the dollar's role as the reserve currency in the world, in which case America would have to pay its foreign debts. And the US economy is already looking feeble.

Even without this defeat, the old idea that you can't fight back and win is disappearing. The oil companies and Washington are weaker. It looks like the space for climate politics, and for all opposition politics, is growing.

The changing movement

Movements change as they grow. They wax and wane. Activists learn. As the movements grow they have the confidence to make new demands and see the need for new forms of organisation. And they can also run into sand.

By early 2008 the anti-capitalist movement was clearly different from the days of Seattle and Genoa. The demonstrations at summits had almost ceased, and the social forums were withering. Marchers had begun by protesting. But as the movement grew people increasingly wanted to really change things.

Mostly that meant changing government policies. Many activists turned toward political parties and elections. The critique of neoliberalism was now the common sense of many. Populations as a whole were moving left.

The most militant and strongest movements were in Latin America. There activists with roots in the new movements began to win national elections. They spoke the language of social justice and sometimes were against imperialism. The leading figures were Hugo Chavez in Venezuela and Evo Morales in Bolivia. The left also won elections in Brazil, Argentina, Uruguay, Ecuador, Peru, Nicaragua, Paraguay and very nearly in Mexico. In Europe the new Left Party in Germany has substantial parliamentary representation and was pulling the whole of official politics to the left. The US saw the same turn to politics, as a whole generation of activists turned from protest to support for Obama.

The activist turn to elections has two sides, however. In Brazil the unions and the environmentalists supported Lula, and in return he backed neoliberalism and allowed the destruction of the Amazon rainforest. Similar compromises happened in much of Latin America, though the governments in Bolivia and Venezuela were much more radical. In the US a Democratic Congress was elected in 2006 with a promise to do something about the Iraq War, and has done nothing. There were no national marches about the war or climate

Climate Politics

change in the US in 2008—activists felt it would not help the Democrats. If Obama is elected, all the indications are that he will try to rein in militancy. In Italy and India too the anti-capitalist left went into coalition governments and found themselves defending neoliberalism and war.

Moreover, the fact that the voters were moving left did not necessarily mean that the left would win. Above all else, people are voting against the bastards who are ruining their lives. Where a neoliberal "left" government is in power, and an angry right wing candidate promises "change", they too can and do win elections.

So the turn to elections can demoralise activists who see their hopes betrayed. But this is only part of the story. For populations are still moving left. The rejection of neoliberalism is spreading and gathering force. Economic hard times, hunger, strikes, war and the effects of climate change are not going to go away. Activists are still grappling with how to move from protest to power.

US activists, for instance, may be thronging to the Democrats because they offer hope of change. But the reason the Democratic candidates have talked so left is that their pollsters tell them the electorate has had it with the way the country is run.

You can never be sure, but my judgement is that over the next few years the tide of opposition will run stronger across the world, and people will look for new ways to change the world. Crucial, though, will be activists prepared to move beyond what the mainstream politicians can imagine. For a climate movement that simply ties itself to politicians in office will rapidly find itself defending compromises that can't save the planet. What we need to do is build movements and political parties and movements committed to winning majorities for radical change, not compromise with the corporations.

In any case, by 2007 the anti-capitalist movement had already changed the way people could think about climate change.

Now

Now suddenly global warming is everywhere in the press. Al Gore's film has been seen all over the world. Politicians of the left, centre and right are tripping over each other to say that something must be done.

One reason the global debate about climate has changed is that the scientific picture is getting worse quickly. This is partly because much more research is being done, so scientists are constantly finding new bad news. It's also because the things they were already measuring are looking worse too.

It now looks like we will need much deeper cuts in CO_2 emissions than scientists assumed ten years ago. By 2001 there was a scientific consensus that abrupt climate change was a threat. By 2007 this consensus was common property among all scientists, and many politicians. The European Union has accepted a goal of limiting total temperature rise to 2.0°C in order to avoid abrupt climate change.

As I explained in Chapter 1, by 2001 it was generally accepted among climate scientists that to avoid a rise of 2.0°C it might be necessary to limit the total of CO_2 in the air to 450 parts per million. By 2007 many scientists were saying that it would probably be necessary to limit the total to well below that, at 400 parts per million.[12] We are already at over 385 parts per million, and the total is rising by 2.1 parts per million a year. But the European Union and the UK government, even in their most radical and abstract moments, are only talking about limiting the total to 550 parts per million, well over either 400 or 450 parts per million.

Even leaving aside the US, the countries of the European Union which have signed up are not on course to meet the limits on CO_2 emissions they agreed to have in place during the years 2008 to 2012. In 2007 global emissions were rising.

So there is now a general consensus among scientists that we are close to a very risky situation and that current government actions are not remotely enough to avoid abrupt climate change. Moreover, each year the outlook appears worse than it did the year before.

People can feel the heat

The third reason politicians are tripping over themselves to say something must be done is that ordinary people can see and feel the effects of climate change *now*.

In 2004 almost everyone talked of the threat of climate change in the future. Activists like me commonly ended speeches by saying we had to act now for the sake of our grandchildren. I did it myself. I don't do that now. The average age of most audiences I speak to is 25. Abrupt climate change will happen in their lifetimes. I am starting to worry it will happen in mine. I'm 59.

People can now see global warming happening on their TV screens and they can feel it in their bodies. Globally, the decisive moment was Hurricane Katrina in 2005. When the hurricane hit, the American media did not mention climate change. Only two years later most people accepted that Katrina was one vision of global warming.

A heat wave in the summer of 2003 killed at least 50,000 people in Western Europe. In 2005 intense rainfall led to floods across central Europe. In July 2007 another heat wave saw record temperatures across the Balkans. That month people in Athens could look at the mountains to the north of the city and see the forest fires raging out of control.

In Australia a long drought had led to marked falls in crop yields, and forest fires across the country. Another long drought was affecting the south western United States. Anyone who walked alongside a glacier in the Alps, the Andes or the Himalayas could look down a long distance and see how far the glacier had melted. By 2006 ski resorts across the Alps were opening late and closing early, and locals were worrying whether they had a future in tourism.

These were the big themes. But the small things were just as important. It was hotter. In March of 2007 I came home to my flat in London one late afternoon. My flatmate Nicola was sitting in the open window of the sitting room reading a book, the sun pouring in around her. Nicola said it was frightening. I asked why and she said, "It's global warming, isn't it?"

That spring I sat with my cousin Winthrop on the lawn under his old oak tree, looking out over Waquoit Bay on Cape Cod in Massachusetts. I said how beautiful it was. Win told me how the next hurricane would take the old oak tree, the bluff it grew on and the land halfway across his lawn to his house. Win and the men's breakfast group at his church had been reading up on global warming.

When the floods arrived in England that summer, everyone—taxi drivers, friends, my colleagues at work—said, "It's global warming."

Everyone knows.

Centre right politicians

Another reason climate politics is changing is that ruling classes and corporate leaders are increasingly worried. After all, as I said before, they own the world. They do not want to be responsible for horror or see their wealth destroyed. Politicians and journalists of the centre and the right have also begun to talk about the threat of anarchy and social chaos that climate change would provoke. "Anarchy" and "chaos" are codes for popular revolts against corporate power. Such revolts are entirely likely once people start dying and suffering in very large numbers. There is also the threat of economic chaos, and of simply losing things they hold dear, like New York and London.

And there is a more immediate worry. In the spring of 2005 a delegation of leaders of industry requested a meeting with Tony Blair before the G8 summit in Scotland. That delegation included the CEOs of several banks, oil, gas and power companies. They told Blair that they wanted him to bring in tighter limits on CO_2 emissions as soon as possible. Their reasons were financial. Many of the present power stations are nearing the end of their useful lives. The business leaders and bankers also expected increasing demand for energy in the future. They were therefore about to decide on investment in a new generation of power stations. However, power stations take years to build, are enormously expensive and last 40 years. The corporate leaders and bankers were reading the scientific journals and the press. They were pretty sure the British government would be compelled to bring in stringent controls on CO_2 emissions at some point in the next 20 years. If they went ahead and built power stations now, under current regulations, when the new regulations came in they would be left with very large useless "stranded" investments.

So the corporations wanted to invest in 2005, but dared not do so without knowing what the regulations would be in ten years time. No individual company could take the financial risk of building their new plants to a much higher standard than their competitors. They needed tight government regulation on emissions now, so they could talk their way out of much tighter controls in the future.[13]

The scientific news is worse. People can feel what is going on. The ruling class is increasingly worried. There is more political space for dissent. All these processes are coming together and influencing each other. The fact that the ruling class is worried makes it easier for the media to spread the news. That makes it easier for scientists to reach people, and easier for people to know what their bodies are telling them. The issue of global warming has now passed a critical point. It cannot be put back in the bottle.

The fact that politicians are talking about climate seriously is also important. It is striking that most of the politicians staking out territory on climate come from the centre and centre-right. The leading voices in 2007 were Al Gore, Arnold Schwarzenegger, Nicolas Sarkozy and Angela Merkel.

Climate Politics

This is not what many on the left expected. There are three important things to understand about the neoliberal politicians and climate. First, they are talking about climate because the majority in the ruling class wants to do something. Second, their climate face is their friendly face in a world where voters increasingly hate neoliberalism. Third, their neoliberalism means that they can talk the climate talk, but they can't walk it. The examples of Merkel, Blair and Gore will make these points clear.

One reason neoliberal politicians are talking about climate is that they are close to worried corporate leaders. They are giving voice to a real debate in the ruling classes about what to do. But they also have specific electoral reasons for becoming climate advocates. There is a quite general move to the left among the general population in most parts of the world. Smart politicians of the centre and the right are looking for ways to make themselves appear as representatives of the radical new mood among the voters. Many feel that the environment and climate are issues where they can stake out such a position. After all, much environmental rhetoric sounds welcoming to the centre and right. Many greens—not most, but many—talk of the necessity of cross-party agreements and working in partnership with business. Making noises about climate also allows centre-right politicians to distance themselves from the globally hated George Bush.

Moreover, neoliberal politicians know that radical positions on pensions and schools bring them rapidly into conflict with business. This is a problem, because they are the politicians who represent big business. But a radical position on climate feels safe. They think it won't mean challenging the corporations. On this they are wrong.

Angela Merkel, for instance, is probably sincere about climate change. She is a physicist by training, and was Germany's environment minister and chief negotiator in the talks that produced the Kyoto treaty. So Merkel knows the score. Since 2007 she has been chancellor (prime minister) of a government coalition between her own right wing Christian Democrats and the somewhat more left Social Democrats. Merkel has tried to project herself as a unifying leader of the nation. Serious talk about climate is the left face of this compromise. Moreover, Germany is one of the few countries likely to make its promised cuts in Kyoto. This is partly because industry collapsed in eastern Germany after 1989. It is also partly because of the influence of the Green Party. Germany now has 250,000 workers in the renewable energy industry and stands to become a leading power in a new energy economy.

In the lead up to the G8 summit meeting in Germany in June 2007 Merkel faced a problem. The most contentious global issue was clearly the war in Iraq. But Merkel's coalition had troops in Afghanistan. Conflict with Bush at the G8 summit over Iraq would make the meeting bitter and strengthen those in Germany opposed to Merkel's policies. Merkel announced that the summit was not about the war, but about climate change. At the summit itself Merkel led an attempt to force Bush to agree to some action on climate. The difficulty was that the Bush administration would not play ball. In the end a form of words was cobbled together saying that the US government would be part of doing some unspecified thing at some point in the future. Merkel was discovering that climate politics was not that much of an easy option after all.

At the same time, Merkel was heavily backing new European Union regulations to limit gas mileage in cars in order to cut emissions. But Daimler Chrysler and Volkswagen are the two largest corporations in Germany. They vetoed the new mileage regulations, and Merkel had to bow to their power. So she began to talk about promoting carbon markets instead.

Blair and New Labour

The politicians of the centre and the right did the same in Britain. The British example is particularly instructive in showing the limits of neoliberal climate solutions.

As an antidote to popular bitterness over the Iraq War, Blair also tried to maintain a caring, human face. In the early 1990s the vehicle for this was a very public concern about poverty in Africa. In 2005 the G8 summit meeting was in Scotland, and Blair faced the prospect of another monster demonstration, this time in Edinburgh, organised by the churches and NGOs in the Make Poverty History campaign. Blair and Brown were able to make an alliance with Make Poverty History. They promised to deliver action on poverty from the G8. In return the churches and NGOs agreed to mute their criticisms, and not to make the Iraq War a theme of the demonstrations.

Alongside the march of 250,000 in Edinburgh, Bob Geldof and Bono organised a simultaneous stadium concerts across the world for Make Poverty History. The acts at those concerts were forbidden to make political statements or criticise the world leaders. When the concerts finished, Geldof and Bono congratulated the G8 leaders on a job well done. At a counter-conference in Edinburgh the next day a furious Bianca Jagger told a cheering crowd that she was a celebrity too, but the difference between her and Bob Geldof was that she didn't sleep with the enemy. In the next two years the G8 leaders did nothing and Africa stayed poor. Make Poverty History was wound up, a shameful embarrassment for all concerned.

Blair's human face turned to climate change. He told the media that his personal project was to convert George Bush. Friends of the Earth, an environmental NGO in Britain, brilliantly called Blair's bluff with a campaign called the "Big Ask". The Big Ask asked members of parliament to support a draft bill providing for 60 percent reductions in UK emissions by 2050. It also called, critically, for 3 percent cuts each year, and proposed ways to make sure those cuts were honoured. Within a year over 400 of the 646 MPs had signed the draft bill.

Facing the possible introduction of such a bill, the New Labour government brought forth their own draft version, and promised to enact it in 2008. They followed much of the Friends of the Earth draft, but made two crucial changes. One was that the cuts became "targets"—that is, nothing would happen if they were not met. The other was that the "targets" would not be annual, but over five-year periods, and the first one would not fall due until 2013. In effect, the bill was now a statement of good wishes. British people had seen scores of New Labour targets not met already.

Friends of the Earth were cautious, but also pleased. Something was better than nothing.

David Cameron, since becoming leader of the Conservative opposition in 2006, has decided to stake out a position to the left of Blair. His strategy was to woo an electorate who were fed up with Blair and New Labour, neoliberalism and the Iraq War, but still loathed the memory of Margaret Thatcher's Conservative governments. To capture those votes, Cameron had to moderate his right wing party, and where possible attack New Labour from the left. Unfortunately, Cameron and the Conservatives wholeheartedly supported privatisation and cuts in government expenditure. They felt the Iraq was a failure, but did not want to call for the withdrawal of the troops. That left climate change.

Cameron flew to the Arctic for climate photo ops in the snow. He bought himself a micro wind turbine, fitted it to his roof and then had it taken down, for all the obvious reasons. He cycled to work one day a week while the security people followed in a car with his papers. And when the draft New Labour climate change bill was introduced in parliament, Cameron got up and called for annual targets, with withering scorn for New Labour's hypocrisy.

As I write, the climate bill is still making its slow way through parliament. At the same time all the other relevant New Labour policies are increasing CO_2 emissions. During the 1990s, under Conservative and then New Labour governments, Britain saw a dramatic fall in CO_2 emissions. The reason was that the Conservatives had closed almost all the coal mines to humiliate the miners' union. The power stations switched to North Sea gas, which produced fewer emissions. With the exhaustion of North Sea gas in the 2000s, however, the power stations began to switch back to coal, and emissions began to climb again.

New Labour have been expanding airports, with new runways at Heathrow and Stansted. They have also announced a massive new road building programme. The new prime minister, Gordon Brown, proposed three million new homes, without the changes to building regulations that would reduce emissions. The new transport minister, Ruth Kelly, announced that government subsidies for the railways would be cut in half, and said the cuts would be paid for by rises in ticket prices.

Yet New Labour boasted of their renewable energy policies which centred on wind power. In 2007 they announced that Britain now has 2 billion watts (Gigawatts) of wind energy capacity. They ignored the fact I discussed in Chapter 5 that Germany already has ten times that capacity and that the state of Texas alone has more than the UK—2.7 billion watts. This is hypocrisy on a staggering scale.

Al Gore

The most important representative of these politicians of the centre and right, and the most impressive, is Al Gore. The story of his political life shows both the limits and the possibilities of neoliberal climate politics.

Al Gore was elected vice-president with Bill Clinton in 1992. Clinton and Gore had both been leaders in the successful attempt to move the Democratic Party to the right, in response to 12 years of Republican presidents. The thinking behind this was that the voters as a whole had moved right, and the Democrats should adopt neoliberal economic policies and right wing social policies in response.

So Clinton supported the death penalty and made a point of signing death warrants as Governor of Arkansas even as he was running for president. When Clinton came into office, there were 1,300,000 adults behind bars in America. When he left office, there were

2,000,000. He continued the Republican economic policies of Reagan and the elder Bush. But he added, for the first time in American history, a drive to balance the budget by cutting federal government spending on human need. In 1996 he signed a law passed by Congress, which in the words of Clinton's adviser Bruce Reed, ended "welfare as we know it".

Gore was, if anything, to the right of Clinton on the economy and society. But what made him stand out was his passion for the environment. His book, *Earth in the Balance*, published during the campaign in 1992, was radical. It was also an early statement of the necessity to act quickly on global warming.

As Clinton entered the White House, his economic advisers told him that if he wanted to have the support of Wall Street and American business, there was no way he could meet his campaign promises of social justice. Clinton acquiesced.[14]

At these meetings Gore held out for the one environmental measure he most wanted—a carbon tax, particularly on petrol, to reduce CO_2 emissions. Clinton said no, that was politically impossible. Gore was only vice-president, an office with no power whatsoever until Dick Cheney came to power in 2001. So Gore went silent on the environment.

A central demand of the neoliberal project was to cut social security pensions by reducing benefits and raising the retirement age. The Clinton administration could not do this without the cooperation of Republicans in Congress. Gore, backed by Clinton, agreed to secretly canvass leading congressional Republicans to gain their support for social security "reform". The Republicans had to explain to Gore that social security was the "third rail" of American politics. Like the electric rail in the middle of an underground rail line, touch it and you're dead.

So for seven years Gore went along for the ride on neoliberalism and mostly kept his mouth shut about the environment. The most galling part of that was that he had to sit and watch as the coal and oil industry dominated American delegations to climate negotiations.

Then he seized the moment. The negotiations in Kyoto were deadlocked between the Europeans and the poor countries and the intransigence of the American delegation. Gore flew into Kyoto with less than two days left to the deadline for the end of the meeting. He used every bit of influence and passion he had to cobble together a compromise that all the countries could agree to. Then he flew home. The Clinton administration refused even to try to ratify the treaty, on the grounds that the Senate would never agree to it.

There are two ways of understanding what Gore did at Kyoto. One is that he got the Europeans and the developing countries to agree to a treaty so full of holes that it would never stop climate change. The other view is that his last minute intervention had forced the American delegation to agree to something. It had kept alive the possibility of a global treaty, one that was finally ratified by almost all countries but the US, and one that will make stronger, more effective treaties possible in future. Both ways of looking at Gore's intervention are valid. But from his point of view, starting from a position of great weakness, he had made an intervention in history.

Then in 2000 Gore got his chance to run for president against George Bush. Gore wanted to campaign on the environment. His advisers told him that would make him look like a tree hugger and turn voters off. That was probably untrue—large majorities of Americans in every opinion poll are always in favour of strong action on the environment. But it would certainly have pissed off American business.

Gore followed his advisers' advice and shut up about the environment. It was a mistake. Instead he campaigned as a traditional Democrat, giving speeches about what he would do for the common man and woman—the "middle class", as the working class is usually called by American politicians. The trouble was that Gore's heart was not in it. He was a committed neoliberal, so his speeches were insincere, stilted and wooden. When Gore talks about climate now you can hear his passion, commitment and humanity. In 2000 his opponent, George Bush, was an oilman, a graduate of Yale and Harvard Business School. To many voters Bush's folksy Texas accent sounded sincere as he promised "compassionate conservatism", controls on CO_2 emissions from coal fired power stations and an end to military adventures overseas.

Gore won the election, but only barely. His margin was so small it opened a space for the Republicans to steal the election in Florida. Even then he could have summoned a movement on the streets in Florida. The Republicans were sending crowds of hundreds of staffers and supporters charging into the recounts. A mass mobilisation for democracy would have cowed the Republicans and the judges on the Supreme Court, and given Gore power. He could not summon that movement. He had been too long in thrall to Clinton, to his own advisers and to neoliberalism. He let the chance go.

And there his story might have ended. But freed from all the constraints of office, he pulled himself together and began to go round

Climate Politics

the world giving hundreds of slide shows on climate change. Most of these shows were for businessmen and "opinion leaders". Gore finally sounded dedicated and honest.

Then, finally, in 2005 his slide show went global with his film *An Inconvenient Truth*. The fact that it was Gore saying these things meant that it got into cinemas all over the world and could not be ignored. And if you did not ignore it, you had to agree with it. People left the cinema visibly shaken. In 2006 the film won the Oscar for best documentary, a choice that was also a political statement by the voters in the Academy of Motion Pictures Arts and Sciences.

People started talking about Al Gore for president. He considered it. But he also told reporters that he was a lot happier not being in politics, that he could do things he couldn't do in politics and that he felt more at home with himself. By and large the reporters didn't understand what he meant.

With the confidence that came from the reception from his film, Gore decided to push his crusade up a level. He organised simultaneous global concerts to promote awareness of climate change. Leading pop stars participated in "Live Earth" in New York, London, Jo'burg, Rio, Shanghai, Tokyo, Sydney and Hamburg on 7 July 2007. The organisers claimed a television audience of two billion.

It was easy to criticise Live Earth, and radical environmentalists and climate activists did so. The pop stars had flown to the gigs in private jets emitting CO_2. The organisers announced that they had made the whole event "climate neutral" by buying "carbon offsets". The activists pointed out that offsets were a scam (an issue I return to later.)

At most of the concerts there was little politics from the stage. The activists said that celebrities would not change the world. They pointed to the experience of Bob Geldorf's Live Aid concerts in 2005, the model for Live Earth. The activists also pointed out that there were now three bills in Congress for action on climate change, and Gore supported the weakest of the three. He was involved with companies which were determined to make money out of climate change action. And the measures he was calling for were still anodyne.

All of this was true. But it missed an enormous point. Gore had broken through the wall of silence in the American media. More than any other person, he had changed the terms of global debate. His personal crusade was the most visible and most supported part of a new movement to stop global warming. And he had opened up a space where it was possible for other people to be far more radical.

Climate Politics after 2001

I said that there was little politics from the stage at Live Earth. But there was also, little reported, an extraordinary speech from the stage at the New York concert by Robert F Kennedy Jr. Kennedy is the eldest son of Robert F Kennedy, murdered in 1968 as he ran for president as an anti Vietnam war candidate. Robert F Kennedy Jr had not gone into politics directly like many of his family. He became an environmental lawyer and leader of River Keepers, an organisation to protect American waters from pollution.

Kennedy had written *Crimes Against Nature*, a savage attack on the Bush administration's environmental record.[15] Kennedy, however, was still part of the ruling class. So his book emphasised the importance of working with business and Republicans, and cited the environmental commitment of his cousin's husband, Arnold Schwarzenegger. But on 7 July 2007 Kennedy stood on the stage in New York, looked out over the crowd, and called for insurrection.

It is worth quoting quite a bit of his speech to show the political space Gore had helped make possible:

Now we've all heard the oil industry and the coal industry and their indentured servants in the political process telling us that global climate stability is a luxury that we can't afford. That we have to choose now between economic prosperity on the one hand and environmental protection on the other. And that is a false choice.

In 100 percent of the situations, good environmental policy is identical to good economic policy—if we want to measure our economy, and this is how we ought to be measuring it, based upon how it produces jobs and the dignity of jobs over the generations, how it preserves the values of the assets of our community and how it averts the catastrophe of global warming.

If, on the other hand, we want to do what they've been urging us to do on Capitol Hill, which is to treat the planet as if it were a business in liquidation, convert our natural resources to cash as quickly as possible, [and] have a few years of pollution based prosperity... But our children are going to pay for our joyride... Climate change is upon us. Its impacts are going to be catastrophic and we are causing it. The good news is, we have the scientific and technological capacity to avert its most catastrophic impacts. We only need the political will...

Now you've heard today a lot of people say that there are many little things that you all can do today to avert climate change on your own. But I will tell you this, it is more important than buying compact fluorescent light bulbs or than buying a fuel efficient automobile.

Climate Politics

The most important thing you can do is to get involved in the political process and get rid of all of these rotten politicians that we have in Washington, DC, who are nothing more than corporate toadies for companies like Exxon and Southern Company, these villainous companies that consistently put their private financial interest ahead of American interest and ahead of the interest of all of humanity. This is treason and we need to start treating them now as traitors...

And I want you to remember this, that we are not protecting the environment for the sake of the fishes and the birds, we are protecting it because nature is the infrastructure of our communities... The air we breathe, the water we drink, the wildlife, the public lands, the things that connect us to our past, to our history, that provide context to our communities and that are the source, ultimately, of our values and our virtues and our character as a people and the future of our children.

And I will see all of you on the barricades.[16]

Global demonstrations

The changed political climate was not changing mainstream politics. A loose global movement of climate protests also began from 2005 onwards.[17]

International climate talks happen every year under the auspices of the UN. They are supposed to discuss the treaty to replace Kyoto in 2012. So far they have made little progress. The campaign chose the talks as an occasion for the global demonstrations partly because they provided a hook to hang the event on. But they were also selected because we all knew that in the end a global treaty will be required to stop climate change. The protesters all demanded that the USA sign Kyoto, and that a strong and effective international treaty follow Kyoto.

The UN talks in 2005 were in Montreal in November. The British campaign began contacting activists, whoever we could, in other countries. We said to them this is only the beginning. The demonstrations are going to be small in most countries, and tiny in many. But if you only get 50 people to demonstrate in your country, that will still be the largest climate demonstration ever in your country. We are establishing a principle here. It will take a mass grassroots movement to stop climate change, and that movement must be global, so join us.

When November came, we had demonstrations in 21 countries—Finland, Russia, Croatia, Greece, Turkey, Poland, Bulgaria, Romania, Norway, France, England, Scotland, Portugal, Mexico, the United

States, Canada, South Africa, Bangladesh, South Korea, Australia and New Zealand.

The organisers of the demonstrations differed greatly from one country to another. In Britain, for instance, the core of the campaign was activists in local branches of Friends of the Earth, Greenpeace and the Green Party and socialists. As support for the demonstrations grew, they were joined by activists from Christian and Muslim environmental campaigns, and by young people already politicised by the social movements.

What was happening was that the campaign had begun on the margins, with an environmentalist understanding of the world but a social movement model of organising a united campaign. That approach brought in people from the political parties which had done little about the environment, and the environmental NGOs which were not used to marching.

Internationally the process was similar. The groups involved and the background of the activists varied widely. But what was noticeable was that because climate demonstrations were a new idea, it seemed that almost anyone could and did start the ball rolling, and then more established groups got involved or spoke at the demonstrations. One thing that was noticeable, though, was that the demonstrations were larger if they were organised by a coalition than if they were organised by one party or NGO.

On 3 November 2005 we had demonstrations in 20 countries. From Zagreb a happy organiser said they were only 70 people and two dogs and the police, but the national media were there, and it was the biggest climate demo ever in Croatia. In Turkey the organisers reported 3,000 in six cities. In London we had 10,000, to our delight. Compared to the Stop the War demonstrations, it was tiny. But it was ten times the size of our largest climate demonstration before. In Australia there were 20,000 spread across all the major cities, and the organisers were laughing with exhaustion.

The largest march was in Montreal, the site of the UN climate talks. There the organisers were hoping for 10,000 people. They got 30,000. John Bell, a Canadian socialist and environmentalist, stood at the metro stop and watched the people coming and coming. He stood in the cold, happy, noticing two things. One was that the demonstrators didn't look like he expected environmentalists to look. They looked like working Montreal, in families, bundled up in parkas. The other was that they did not come as organisations, and any signs or banners they carried were home made. Later I called the organisers. "Thirty thousand people," the voice of someone I didn't

Climate Politics

know said down the phone. "And we don't know who they are," the man said. "And they're shouting too much for me to hear you. You want to hear them shout, Jonathan?" He held his phone out and I heard them shouting.

The Montreal demonstration came in the middle of the two weeks of the UN conference, and made a difference. The United States delegation, on orders from Washington, left the conference, in an attempt to kill it stone dead. The then Canadian prime minister had seen the marchers, though. He stepped in to support the talks, and defy Bush, by asking Bill Clinton to speak. The US delegation marched back into the conference, and the endless global negotiations were still, tenuously, back on track.

By the next year, 2006, we had a very loosely organised Global Climate Campaign and demonstrations or actions on 3 December in Bermuda, Bolivia, Brazil, Colombia, Panama, Bermuda, Australia, New Zealand, South Korea, Taiwan, Bangladesh, USA, Canada, South Africa, Kenya, Nigeria, Turkey, Russia, Finland, Sweden, Denmark, Norway, Bulgaria, Serbia, Croatia, Slovenia, Greece, the Czech Republic, Italy, France, Germany, Belgium, the Netherlands, Portugal, Britain and Ireland.

Again most of the demonstrations were small. But the demonstrations were spreading to the global south, with solid organisations in several Latin American countries and in Bangladesh. 5,000 Kenyans marched in Nairobi, where the UN talks were held. And in Europe the NGOs were becoming more involved on the ground.

The largest turnout was in Australia. The right wing government there was the only other developed country that had joined the US in refusing to sign Kyoto. So 40,000 marched in Sydney, 30,000 in Melbourne and another 20,000 in other cities. This was large enough for Australian climate activists to feel they could make a real difference to national politics. For 2007 they set the date for their national demonstration just before the national elections, figuring that might help tip the balance, prevent the re-election of prime minister John Howard and change national climate policy. 130,000 people marched, Howard lost and the new Labour government signed Kyoto immediately.

There were more global demonstrations in December 2007. None of them was anywhere near as large as the Australians', but people protested in over 70 countries, including 11 in Africa and eight in Asia.

In 2005 and 2006 the protests in the United States covered more than 20 states, but were generally small. In 2007 this changed. In early January, Step It Up set up a website and asked people to organise marches and rallies in their cities on 14 April.[18] Step It Up was basically a respected environmental writer, Bill McKibben, and some students at Middlebury, a small liberal arts college in Vermont. Within three months people had written into the website from 1,400 towns and cities. The organisers received reports from protests in over 800 of those cities by roughly 150,000 people. The demand was for the new Democratic Congress to pass a law to reduce CO_2 emissions. All the protests carried one simple banner or placard—"80 percent by 2050—Congress: Do It."

The Step It Up protests showed that feeling in the US, as globally, is beginning to swell. They also showed, as the demonstrations had in other countries, that the first impetus for action can and does come from unexpected places. In a political situation where the major established organisations are slow to act, people rally to whoever is calling for action.

Still the marches everywhere are nowhere near the sort of mobilisations we will need to begin to force governments to halt climate change. But they are a beginning. Environmentalists are turning to try to organise mass movements to defend the planet. Organising a demonstration means joining the ranks of protesters. It is a break from the old ways of lobbying the top and trying to work through the media. It is happening because so many environmentalists are changing, and many ordinary citizens are deciding that something must be done.

Demonstrations on their own do not change the world. But they do change the demonstrators. I have said before that one of the central barriers to action is all the people who say they want to act, but no one else does. If we get each person who says that on the streets, marching with the other people who say that, then no one will feel alone. If demonstrations, whatever form they take, grow large enough, activists all over the world will take heart. Then they'll go home from the marches to petition, to lobby politicians, to elect radical parties, to press employers, to blockade and to do whatever else it takes.

So demonstrations will not be the movement. But they are a way to start a movement. The next chapter looks at the debates now raging over what a global climate movement should fight for.

Climate Politics

CHAPTER 16

Personal and Market Solutions

To make sense of the current global debate about climate change, we have to start by recognising that the contradictions of mainstream climate politics have not gone away. The carbon corporations are still there, still powerful and still working flat out to prevent any real action. This must never be forgotten. But globally the mainstream of corporate leaders and neoliberal politicians want to do something and are under increasing public pressure to be seen be doing something. They fear that, if they do not act, their current investments in power stations and carbon energy will be wasted when an angry public finally demands strict controls. But their allegiance to the market sets strict limits to what they can actually do.

The corporations and the neoliberal politicians are pushing two kinds of solution to the problem—personal consumer choices and market incentives. In this chapter I explain, in more detail than previously in the book, why the corporations and governments are promoting these solutions and why they won't work. Personal solutions fit with the ideas of the neoliberalism. They make sense to people who want to preserve the power of the market. And they make it appear that the governments and corporations are doing *something*. For all these reasons, personal solutions are a powerful distraction from effective action.

Carbon footprints

Every week local groups in Britain and elsewhere come together to try to do something about global warming. At the first couple of meetings, these groups discuss a simple question—what do we do? People in that room usually have two different answers to that question. One is political action to press for change in government action and cuts by the big corporations and employers. The other answer is individual action.

The first step is to go to a website or buy a book that helps work out your personal "carbon footprint"—the sum of the emissions your personal consumption is responsible for. So you add up the emissions from your car travel and air travel. Your gas, electricity and heating bills tell you how much you emit at home. Then you add up your main spending on manufactured goods and imported food that costs "air miles" to reach you.

That gives you your total carbon footprint—perhaps ten tons of CO_2 a year. Armed with that, you work out ways to reduce your footprint. You can, for instance, ride a bicycle, take the bus, cut home energy use, buy better light bulbs, turn off the computer at the wall, buy solar panels for the roof and eat only locally produced food.

The flaw in these personal solutions is that they start in the wrong place. Environmentalists have a traditional phrase for this. They say that personal solutions start from "the wrong end of the pipe".

The idea of the end of the pipe comes from many campaigns against factories that put poisons into the rivers. Environmentalists have learned that there are two strategies to stop that kind of pollution. One is to get the government or the courts to forbid the factory to use that poison. That stops the pollution before it even gets into the pipe.

The other strategy is to try to deal with the pollution at the other end of the pipe, after it is already in the river. At that end, you can campaign for local government to spend more cleaning up the river. You can monitor the level of poison coming out of the pipe. You can sue the polluter and win compensation for the damages they have caused. You can tell everyone to drink bottled water. Or you can move.

Environmentalists have learned that it's best to stop the pollution before it goes into the pipe. Personal consumer choices to cut CO_2 emissions start at the wrong end of the pipe.

The most important changes we need can't come from individual consumer behaviour. Wind power needs giant wind farms. A few people can go solar, but it will take a government programme for everyone to go solar. Some people can be persuaded to ride a bicycle. But if cars are banned and good public transport is provided, the carbon savings will be far greater. Individuals can't shift freight to rail. They can't regulate industrial processes. They won't insulate all the old buildings, and most of them can't afford to buy new passive houses.

On some level, everyone knows this. So when there is a debate in the local climate group about what to do, most people in the room sort of agree with those who say they have to campaign for political

Climate Politics

solutions. But then someone says that's all very well, but what are we going to do? And then the group starts talking about reducing everyone's carbon footprint and gets lost in the detail.

Behind this lies a political fact. Many of the people in the room think political action is a good idea. But they don't think it will work. They don't think ordinary people can change government policy over something so important. In a sense, they are both only too aware of, and in awe of, neoliberal policies and corporate power. Tackling the powers that be seems nearly impossible. Yet because the threat of global warming is dire and immediate they want to do *something, now*.

Individual actions may only make a small difference, but they do have one great virtue. They are an act of witness. If you ride a bicycle to work or don't take the plane, your action provokes hundreds of conversations with friends, family and workmates. All those conversations are about global warming and how something must be done about it. But if these discussions lead people to think that personal choices are the main solution to global warming, we will fail to stop the threat.

Moreover, when you look at your carbon feet, you should also look over your shoulder to see who is promoting this solution. There is now a relentless campaign to tell us that individual solutions are the answer. In Britain these days the media constantly ask what people are doing about global warming—and the answer always focuses on your personal consumption. The great corporations, including the oil companies, run ads in the papers and on television encouraging people to examine their consciences. The same governments who do nothing also encourage climate education in schools that teaches children to blame themselves and their parents, not the government. A brief history of garbage in the United States is the best way to illustrate this.

Garbage

Heather Rogers, in her wonderful book *Gone Tomorrow: The Hidden Life of Garbage*, has given two historical examples of how corporations have promoted action at the wrong end of the pipe in the United States.[19] Both should make us think hard about individual solutions to global warming.

Rogers' first example is the invention of the idea of "litter". In the US in the 1950s industry was producing increasing amounts of packaging and disposable products. The packaging was justified in marketing terms—people would be more likely to buy the main product. But the real beauty of disposable packaging for manufacturers was that the packaging itself had to be thrown away each time. Three inventions were key—plastic wrapping, the disposable plastic bottle and the disposable tin can. Until they were invented, beer and soft drink bottles always had to be returned to the store, then to the company, then washed and reused. The bottle company was only paid once. Now the can could be thrown away each time, and so the can manufacturer could sell far more cans and the plastics manufacturer far more wrapping. Industry was expanding in the 1950s and looking for new markets. Before disposables breweries and soft drink companies had local bottling plants. But with disposable bottles and cans, most plants closed down and there was a great concentration of companies in the industry. The cost of packaging rose to $25 billion a year, a substantial part of the output of American industry.

All this plastic packaging, and all the bottles and cans, counted as economic growth. But they were waste, of no use to anyone. They rapidly began to cover the landscape and to fill the landfills, at considerable cost to local governments. State legislatures began to talk about banning packaging, and Vermont actually banned throwaway bottles.

The packaging and soft drink industry reacted by launching a new campaigning body, Keep America Beautiful (KAB).

According to Rogers,

[KAB's] founders were the powerful American Can Company and Owens-Illinois Glass Company, inventors of the one way can and bottle, respectively. They linked up with more than 20 other industry heavies, including Coca-Cola, the Dixie Cup Company, Richfield Oil Corporation… and the National Association of Manufacturers, with whom KAB shared members, leaders and interests…KAB came out swinging, urgently funnelling vast resources into a nationwide, media-savvy campaign to address the rising swells of trash through public education focused on individual bad habits and laws that steered clear of regulating industry.[20]

Keep America Beautiful discovered and popularised a little used word—"litter". And they invented a new word—"litterbug"—for the evil person who causes the litter. They ran ads with pictures of little monster litterbugs dropping litter. They promoted anti-litter

Climate Politics

laws, with jail terms for repeat offenders. And they persuaded most Americans that the problem was not packaging, but individuals who did not use the packaging right. Vermont dropped its law against throwaway bottles. In the words of one executive at the American Can Company, "Packages don't litter; people do".[21]

For a time the packaging industry was sitting pretty. But then a new wave of environmental consciousness swept the US in 1970. Keep America Beautiful was still very much in business:

> On the second Earth Day in 1971, [KAB] premiered the first of its now iconic television advertisements starting the buckskin-clad long-time Hollywood actor Iron Eyes Cody... The haunting ad was seared into the guilty consciences of Americans young and old: After stoically canoeing through a wrapper-and-can-strewn delta, past a silhouetted factory puffing smoke, Cody dragged his canoe onto a bank sprinkled with litter. Hiking to the edge of a freeway clogged with cars, the stereotyped North American was abruptly hit on the moccasins with a fast-food bag tossed out the window by a free-wheeling blond passenger. Cody then looked straight into the camera as he shed a single tear. The accompanying music was stirring, the voice-over solemn: "Some people have a deep, abiding respect for the natural beauty that was once this country. But some people don't. People start pollution. People can stop it".[22]

The collective problem had been rebranded as individual failure.

This approach had considerable success, but did not carry all before it. In the 1970s and 80s the environmental movement began to get approval for local laws about packaging. Governments were also finding it harder to dispose of waste. The price of landfill garbage had stayed pretty steady for 30 years. Then new federal health and safety regulations governing landfill sites more than doubled the cost of burying waste between 1984 and 1988. The 1980s also saw many successful community campaigns against garbage incinerators, which poisoned the surrounding air and water.

The packaging, can and bottle companies could feel the pressure rising. They had to find some way to dispose of their waste. In these circumstances, they embraced recycling. The idea had come from environmentalists. And it is certainly better to recycle than to throw away. But now the corporations embraced it, particularly the Society of Plastics Industries, the American Plastics Council, and Keep American Beautiful. Recycling was not first choice for the polluters, but it did have advantages.

First, recycling allowed the polluters to continue to pollute. It put the responsibility back on the shoulders of the person in the kitchen sorting the trash. Recycling seemed to be doing something about a problem that would never have existed in the first place if it weren't for the packaging. It blamed the individual, and changed the subject.

Second, recycling concealed the reality of waste in America. Recycling was applied to what is called "municipal waste". This is garbage from homes, schools, public buildings, restaurants and hotels. It accounts for less than 2 percent of the waste in America. More than 98 percent is "industrial debris from mining, agriculture, manufacturing and petrochemical production". Another way of saying this is that, for every ton of household and municipal waste, there is more than 70 tons of industrial waste.[23] Some of this industrial waste is recycled, if the company chooses. The great majority of it is not. Industrial waste is also more likely to contain toxic substances.

What this means is that recycling by individuals has a very small effect on American waste. Moreover, large proportions of the waste that are put out for recycling are actually dumped and go to landfills. Most materials can only be recycled once before the fibres break down. Much material for recycling is sent abroad, at great expense of carbon. Recycling is like cutting both your arms and then bandaging the cuts on one arm. It's better than nothing. But really, someone ought to take the knife away.

Can't we do both?

The point about both litter and recycling is that they relocate the problem. The corporations are allowed to continue packaging and polluting. The individual becomes responsible. That means people blame each other, and can feel either smug or guilty. In any case, they take their eye off the ball.

Something very similar is happening with the focus on carbon footprints. So although I have some sympathy with my friends who try to change their lifestyle, I am angry at the politicians and corporations who urge them to do so. My friends start from wanting to do something. The politicians and corporations are making the individual seem responsible.

As I said above, every local group of climate activists faces a choice about what to do, and two possible answers come up every time. They could be a campaigning group to force governments and employers to act, or they can educate people locally about how they can change their lifestyles. In theory, a local group could do both. In practice, actual groups choose to make one more important. The choice is clear in what the first leaflet says. Does it talk about demonstrations or carbon footprints? The same choice lies behind what you say to the person who comes up to your stall in the city centre on Saturday.

Most local groups in Britain currently opt for changing lifestyles. One problem with this is that lifestyle decisions mark people off from majorities. But to stop global warming we are going to have to get the majority of the world's people to act. That means the majority in every large country, at a minimum. An emphasis on personal changes has a built in tendency to exclude these majorities. For one thing, most carbon lifestyle changes cost money and many involve bank loans that some people can't get. Many people recommend changes that assume you own your own house. They assume you can invest money now in solar power and make the money back over the lifetime of the house.

Where I live, long distance trains cost more than cheap flights. It costs less to drive 70 miles to my work than take the train. Hybrid cars cost far more money than I'll ever see. And it's no good telling most people they will save money in the long term. The richest fifth of society save money on many things by planning ahead. Most people live from one pay cheque to the next.

The next difficulty is that people pushing lifestyle changes often appear to be smug and superior. Sometimes, indeed, they are. Most, however, don't want to come across as arrogant, but still feel a strong moral imperative to do something about climate change. I feel that too. But from there it's only a small step to guilt when you emit carbon. It's another small step to thinking other people who don't change their lifestyles are bad. Moral feelings become moralism, a way of making some people better than others. People sense these kinds of judgment, and hate you for it, particularly if you can afford to do the right thing and they can't. They hate you even more if they drive a truck, need the job and already feel bad about the environment.

Individual lifestyle changes leave out most people on ordinary incomes. And they don't apply to the poor countries. No one is seriously proposing that there are individual solutions for India, China and Indonesia. Moreover, most personal carbon choices come along with an ethic of sacrifice. This is built into the project. If you don't go for big society-wide solutions, then people on the ground will have to make sacrifices. If the government does not insulate houses, people will eventually have to turn down the thermostat and freeze to save energy. If governments do not rewire the world, we will all see the rains fail and the crops fail.

So individual lifestyle changes almost all mean paying more or giving up something. This runs into the problem that most working people even in rich countries don't want to sacrifice.

Finally, there is a weakness built into the politics of individual change. It starts from a desire to do something and a fear that ordinary people are not strong enough to act together. Individual action then confirms that fear in two ways. First, in the ordinary process of saying it, you persuade yourself and other people that collective solutions are not possible. But secondly, all the people you fail to persuade confirm your feeling that people are no good.

Having said all this, I don't walk away from those meetings where people talk about their carbon footprints. I know what they want to do. But the method they are choosing can't solve the problem. That's why the corporations and governments are now pouring money into advertisements and education programmes that teach people to blame themselves.

Market fixes

Personal lifestyle changes are not the only answers to global warming politicians and corporations are pushing right now. Their other favourites are market solutions—"green taxes", "cap and trade" carbon markets, "carbon offsets", "green taxes" and "carbon rationing". The rest of this chapter explains what these market mechanisms are, why they won't work, and how they make effective solutions more difficult.

I begin with "green taxes", which is a bit unfair of me. Green taxes are by no means the worst alternative. But they show particularly clearly the difficulty with all market incentives.

The idea of green taxes is simple and attractive. Green taxes force people and companies to make good lifestyle decisions. You put a tax on aviation fuel so plane tickets are more expensive and then people fly less. You tax roads so that people use their cars less. Or you tax cement so people build houses out of something else.

The big problem, though, is that green taxes are still at the wrong end of the pipe. They always have much less effect than simple regulation. Taxing old fashioned light bulbs reduces the number sold.

Climate Politics

Banning them means none are sold. Taxing car parking spaces outside houses reduces the number of cars. Banning cars reduces the lot. Taxing cement could cut some consumption some, regulation could cut most of it. Taxing aviation fuels will never work as effectively as banning all flights within Europe. If a thing is worth doing, it's worth really doing it. Green taxes are always second best.

Green taxes are also always unfair. The first class passenger on an airline uses five times the space of a regular class passenger. Even if the rich man in business class pays five times the extra tax—and he won't—the tax will mean less to him than it does to the woman in the cramped seat. The same applies to taxes on cars in rich countries. It's true with redoubled force of taxes on heating. In Britain thousands of the elderly poor die of the cold every year already because they cannot afford to keep the heating on high.

Green taxes are the opposite of what happened in Britain and America with rationing in World War Two. When food was rationed in Britain, the point was that everyone got the same. People accepted rationing because it was fair. With green taxes, unfairness is built in. If you weren't going to be unfair, you would ban something across the board or ration it equally for everyone. And a strategy which creates and exacerbates inequality leads to a political problem. Sooner or later the more right wing papers will be able to mobilise ordinary people against the environmentalists.

One argument for green taxes is that they can be spent on the environment. For instance, you can use a road tax to pay for public transport. But in actual practice it doesn't work like that. Governments can and do move money around from one budget to another.[24] In any case, governments don't need green taxes to fund public transport. They already subsidise wars and nuclear plants out of general taxation. A government that doesn't already want public transport is not going to spend the road taxes on buses. It will spend it on roads or wars anyway.

One suggestion is that the money from green taxes can be "ring fenced" for environmental spending, and the budget can't be raided. But we have a lot of experience with government "ring fencing". As soon as the government needs the money for something else, it pushes the fence over.

Moreover, in the last 30 years most governments in the world have reduced corporation taxes and income tax on the rich by a great deal. They can get far more money by raising those taxes back to where they were in 1980 than they can by green taxes. If they wanted to spend on the environment, they would do that.

There is a further difficulty. Many green taxes give governments an interest in preserving ungreen behaviour. Something like this already happens with taxes on cigarettes. Cigarette taxes are set to encourage some people to smoke less, but not so high that everyone stops and the government makes no money. In the same way, governments that charge for roads will keep the tax high enough to deter some drivers, but not so high that their tax income falls. And they will not abolish car driving in the city centre.

In short, green taxes work at the wrong end of the pipe, and there is always a better alternative.

Cap and trade

"Cap and trade" carbon trading is another kind of market fix.[25] Ideally, it works like this:

The government of a country decides to cut carbon dioxide emissions by 3 percent a year. So it puts a legal cap, or limit, on the total amount of emissions by all businesses in any one year. Let's say the emissions by all companies were 1,000 million tons tons last year. This year they will be allowed 970 million tons. Next year it will be 940 million tons, and so on. The total of emissions is shrinking each year.

The government calculates how much emissions each company makes at the start of the process. Then it gives each company permits to emit their fair share of the shrinking total. Let's say a power company emitted 10 million tons of CO_2 last year, 1 percent of the total for the whole country. This year they will have permits to emit 3 percent less, or 9.7 million tons. Next year the limit for that power company will be 9.4 million tons, and so on.

But some companies will find it much harder to cut emissions than other companies. This is where the permit system comes in.

Let's say that in the first year our small power company has only 9.7 million permits, but it emits 9.8 million tons. But there will be other companies who have made more cuts in emissions than they had to, and they will have spare permits. So the power company can buy an extra 0.1 million permits, and they're square.

The beauty of this is that there is a financial incentive. Companies can make money by cutting their emissions more than the government asks them to. And the companies who cut most will be the ones who find it cheapest to do so. So all the companies, taken together, will cut the maximum amount of CO_2 at the lowest price.

Climate Politics

Sounds good. Cap and trade schemes are now very popular with neoliberal politicians. The European Union now has a cap and trade scheme as part of the Kyoto Protocol. Most of the new proposals for cutting emissions in the United States involve cap and trade schemes for states and companies. So let's look at how cap and trade actually works.

Market proposals always start by describing an ideal market that does not exist anywhere except in economics textbooks. They then demonstrate how well that ideal market would work. So you always have to look at actually existing markets to see how they work. Otherwise it's like reading romance novels in order to understand your partner.

The idea of carbon trading comes from the Kyoto Protocol. We have already seen how this works. If Britain cannot make the necessary cuts, it can buy credits from Ukraine. Ukraine has the credits spare because its economy crashed in the 1990s.

Fair enough. But if Britain was not allowed to buy credits from Ukraine, Britain would have to cut more. The net effect of carbon credits in Kyoto is that fewer total cuts are made. This is not something special to the Kyoto scheme. It is built into all carbon credit schemes. If the general limit of cuts is 5 percent per country, carbon trading will always make sure that the total cut is never more than 5 percent.

It is important to understand that this is not a case of market mechanisms failing. Some of the American negotiators at Kyoto may have been kidding themselves, and some were blinded by stupidity about markets. But the American business people who insisted on market mechanisms in the Kyoto negotiations knew what they were doing. The example they kept citing was the US cap and trade scheme for sulphur dioxide. It's worth looking at how that scheme actually worked.

In the 1980s governments all over the world realised that power stations were pumping out sulphur dioxide creating sulphuric acid in the air that fell as acid rain and killed forests all over North America and Europe. So the governments, under pressure from environmentalists, put in regulations to control these emissions.

The German scheme to cut sulphur emissions began in 1982 and finished in 1998. Companies were simply told to cut their emissions, and by the end emissions were down 90 percent. In the US the scheme began in 1990. By 2010 US emissions will have been cut by only 35 percent. That's because the US scheme didn't order companies to cut emissions. Instead it operated as a cap and trade market for sulphur emissions.[26]

The people advising and controlling the American delegation to Kyoto were not naïve. They were men from the coal companies, with years of experience on the state and federal level in back rooms with politicians and regulators. The coal company men knew how to read fine print, and what had happened before. If they backed carbon trading, it was because they already knew sulphur dioxide trading was not going to be a problem in the US. After all, most of the sulphur comes from coal.

The EU scheme

The European Union trading scheme provides another example of what's wrong with carbon trading. The scheme began in 2006 and covered only half of European CO_2 emissions. Only the major polluting companies and energy plants were involved. Each national government handed out a certain number of credits to each of their major polluting companies. There was no auction—the companies did not have to pay for these credits. There was no limit on how many permits a government could hand out either. So most handed out more than the companies needed. This wasn't known when the European carbon market started trading permits in 2006, so the initial price was 30 euros for a ton of CO_2 emissions. But as soon as companies realised there were more than enough credits to go round, the price of a permit in the carbon market fell to just half a euro a ton.

The EU department in charge exposed the scam, and made it clear they expected more honesty in future. But there was still no central method of control and no way of taking the offenders to court. Many thought that the EU system would soon have to move to an auction system that would eliminate cheating in handing out permits. However, it would not eliminate cheating in reporting figures at the end of the year. And there are no controls on that.

More important, if companies pay €40 a ton for emissions permits in an auction, they might as well be paying a carbon tax of €40 a ton. That means an auction system would have all the drawbacks of other green taxes. As with taxes, companies would be *encouraged* to cut emissions, but they would not *have* to cut emissions.

The larger problem is that the EU scheme ignores a large proportion of the places where savings can actually be made. It leaves out petrol, all forms of transport, government activity, public buildings and military emissions. Critically, it leaves out households. Yet as we

Climate Politics

have seen, roughly half the total emissions in Europe come from houses and public buildings, and the major solutions are insulation, turning off air conditioning, changing light bulbs and regulating appliances. The EU carbon trading scheme only targets the electricity company supplying power to the building. The supplier can solve part of the problem with wind power. But the supplier can't change the light bulbs and insulate the houses and turn off the air conditioning.

The problem goes deeper, though. There is a reason the German sulphur dioxide scheme has already achieved 90 percent cuts and the American one is unlikely to reach more than 35 percent. The regulation of industrial processes makes companies innovate. Taxing them at levels they can afford encourages them to keep on doing what they were doing. In Germany the result with sulphur dioxide was a lot of industrial innovation in Germany, and in the US almost none.[27]

There are two final points to be made about every cap and trade scheme. One is that they are very hard to understand. The official explanations take hundreds of pages and are littered with jargon and acronyms. Almost no one understands carbon trading schemes. This means that ordinary people can't tell if they are working or not. That, we must understand, is part of the point. It means governments can be seen to be doing something while doing almost nothing.

Finally, people often say that the wonderful thing about carbon markets is that they provide companies with an opportunity to make money out of cutting emissions. They don't. Carbon traders can make money, but someone has to pay their salaries. It's the same as with trading in the market for prices of agricultural goods. You can bet on the future price of pork or corn on the Chicago Exchange. There you can make, or lose, a lot of money. The brokers who handle your bets will make money. But it won't save farmers a penny. It just adds to the price finally charged for pork and corn. It's the same with carbon trading. The company that cuts its emissions has to pay for the new technology, and then it has to pay the salaries of the carbon traders. Some people make money out of the market, but it creates no new money.

Carbon offsets and carbon neutral

There are worse things than carbon trading, however. "Carbon offsets" and "carbon neutral" are worse.

The idea of carbon offsets is that someone has emitted too much carbon dioxide. Let's take a rock concert, with all the bands flying in across oceans on their private jets. The concert promoters are responsible for thousands of tons of CO_2 emissions. Bad karma and bad PR. So they find a tree planting project somewhere in Brazil or Indonesia. A specialist accountant estimates how much CO_2 the tree planting will save. The rock promoter estimates his emissions. Then he pays the tree planting project a sum of money and claims credit for an amount of emissions saved. The promoter then "offsets" the good tree savings against his bad jet emissions. This is a "carbon off-set". The rock concert is now "carbon neutral". It is said not to have added any net carbon dioxide to the atmosphere.

Another kind of carbon offset happens when a guilty traveler returns from holiday. She goes to one of the many carbon offset web-sites. The website helps her calculate how much CO_2 her trip put into the air. She then pays the website a sum to make up for that, and they promise to give it to tree planters or something similar.

Companies do this too, particularly companies under pressure for bad environmental practices. They promise to reduce their emissions by 50 percent. Then they buy carbon offsets. The British government plans to do the same. The climate bill proposed for 2008 calls for 60 percent cuts in emissions in Britain by 2050. However, the bill says that if Britain can't make these cuts, it can buy offsets to make up the difference. The effect is to build into the bill excuses for failure.

Carbon offsets are often scams. Even where they are legitimate, they don't decrease emissions. This takes some explaining, because it takes some believing.

First, there is no regulation of carbon offsets, because there is no legal enforcement. Most happen between countries, where there is not even a shared legal system. So there is nothing that says the company in Brazil actually has to plant those trees. Nor does any-one inspect them to make sure they have. Often they simply don't. The absence of regulation means there is no way of telling what pro-portion of offsets are scams.

Second, even if the Brazilian company does plant trees it will prob-ably be a the kind of scam I outlined earlier—the plantation of fast growing eucalyptus that empties the water table, kills the vegetation underneath, and is cut down in ten years for biofuels or timber. Often this will be on land that was rainforest before. Indeed, many carbon offset companies simply pay people who already own forests for the "carbon offset rights" in those forests. Even where the scam is not that blatant, offset schemes usually involve something that was going to happen anyway. So, for example, a power company in Indonesia wants to convert from a coal fired plant to a gas fired one. They will do this because gas is cheaper for them. But gas produces much less CO_2 than oil. So on the side the Indonesian power company sells the CO_2 savings to a carbon broker. The result is that no actual CO_2 emissions in Indonesia have changed. But the carbon offset has allowed the rock promoter, tourist, oil company, church or city in Europe to emit *more* than they would have otherwise. Remember, the rock promoter or the oil company are buying the offsets because they are under pressure to cut emissions. The offsets are their way of not cutting their emissions and still looking green.

This is important and weird—carbon offset schemes mean more CO_2 emissions than there would be if the schemes did not exist.

The market also creates a constant pressure to make carbon offsets less effective. Ineffective offsets are cheaper. Think about the tourist who comes back from her plane trip and goes to the internet to buy offsets for her emissions. If she has any sense, she will compare the prices on different websites. Which website do you think she will buy her offsets from?

The one with the lowest price? This will be the one trading in scams and fiddles because these create cheaper offsets. Or the company with effective offsets that charges five times as much per ton?

She'll choose the cheaper site. So will the rock promoter, the oil company and the government. This creates a constant pressure towards cheaper credits. Even when the projects are not outright scams, the pressure is always towards projects the company in Brazil or Indonesia would have done anyway.

There are also official UN offsets as part of the Kyoto Protocol. These are called the "Clean Development Mechanism" (CDM). The central problems with CDMs are the same as the problems with all carbon offsets and carbon trading.[28] Companies are encouraged to do something they should be compelled to do. This does not mean that climate aid from rich countries to poor is a bad idea. It is absolutely essential. But it needs to be separated from giving companies excuses not to cut their emissions. What is needed is massive aid to poor countries to fund them building renewable energy, insulating houses, building public transport and building better steel plants. That would make an enormous difference. It could be paid for out of taxes on corporations and the highest earners. It would reduce CO_2 emissions, not increase them. And it would happen in ways that everyone could understand and make sense of.

But that wouldn't be the market, would it? That would be people helping help each other, and is anathema to neoliberalism.

Carbon rationing

The final kind of market incentive now being debated is "carbon rationing". Unlike the other market schemes, the pressure for this does not come from corporations. It comes from environmentalists who genuinely want to stop global warming and support social justice.

The idea of carbon rationing came originally from the thinker Aubrey Mayer, a member of the British Green Party. It is part of Mayer's ideas for "Contraction and Convergence".[29]

Contraction and convergence has been widely praised by many climate campaigners. It starts from a simple idea: every country in the world should eventually have the right to emit the same amount of carbon dioxide per person. This cannot be done quickly. In the meantime, the rich countries will steadily reduce their emissions—contraction. The poorer countries will increase their slightly, or hold steady, until everyone is at the same level, a level of global emissions the world can live with—convergence. This is fair. Most people who support contraction and convergence do so because they think fairness is right, and I agree with them.

However, Mayer's ideas are almost always presented in terms of sacrifice. They are phrased in terms of humanity as a whole giving up goods and services, and the pain being equally spread. I have argued at some length that we cannot and need not build a movement on sacrifice. And there is no reason why the idea of equal emissions for each country has to be phrased in terms of sacrifice.

The second part of Mayer's idea is what he calls "carbon rationing". It is this part that the more radical writers on climate, like Larry Lohman and Ross Gelbspan, have trouble with. Carbon rationing has all the drawbacks of other market incentives.

The idea of carbon rationing is that each person in a country will be given a card like a credit card, connected to a central computer. Everyone will start the year with the same ration of carbon credits. Each time you make a purchase which involves a lot of carbon emissions, it's marked against your card. Examples would be airline tickets, car petrol and electricity bills. Each year everyone's ration would be reduced by 3 percent or 5 percent or 8 percent.

This seems a lot like rationing in the Second World War. And it seems fair, since everyone would get an equal carbon ration. Fairness was the key to public acceptance of rationing in the 1940s. Moreover, fairness was assured because it was illegal to sell your

Climate Politics

ration. There was some trade on the black market, but this was widely regarded as wrong.

The key difference with carbon rationing is that people would be allowed to sell part or all of their allotment for cash. Thus, a London businessman who makes a lot of plane trips to New York could buy up other people's allowances. And a family in Britain who wanted to visit the grandparents in Bangladesh would have to do the same.

It is often said that carbon rationing would introduce a new currency for carbon. But if carbon credits can be sold for dollars and euros, they are not a new kind of money. They are dollars and euros. In effect, they would operate as a kind of tax allowance, and as the mirror image of a green tax. Moreover, year by year people would have smaller carbon allowances, and the price they could sell those credits at would rise. The result would soon be that only the rich could own cars or take planes. The rich could buy credits to heat big houses. The poor could not.

Unlike other market schemes, carbon rationing would reduce the total use of energy. But it would do so by increasing inequality. This form of rationing appeals to most rich people. It will be harder to persuade ordinary people of the benefits.

There is another problem with equal rations of carbon. It goes back to the businessman flying to New York and the family flying to see relatives in Bangladesh. Most people would agree that the businessman could do his work by video link, while the grandmother needs to touch her grandson. The needs are of a very different order. Economic benefits that simply give people the same money are not actually equal. Some people need more help than others. You don't give people equal amounts of money to spend on a national health service. Instead you spend a fortune on people with long term cancers and very little on those who die quietly in their sleep.

Moreover, although people usually need similar basic food rations, they are likely to need very different amounts of carbon. A householder with a 4 kilowatt solar roof needs practically none for heating. The same is true of someone living in a forest who burns logs. An old person living in a large leaky building needs a lot just to keep warm in winter. If you give the same carbon ration to both the person with the solar roof and the old person, you are making the one who is already rich richer.

The biggest problem with carbon rationing, though, is the same as with carbon trading. It tackles the problem at the wrong end of the pipe, with the consumer. Carbon rationing, and contraction and convergence, do not touch the major changes that are needed. Consumer choices cannot cover the world with wind and solar power, or insulate all the houses. Carbon rationing is not a substitute for government regulation and government expenditure.

PART FIVE

ALTERNATIVE FUTURES

CHAPTER 17

Capitalist Disasters

THIS chapter and the next are about two alternative futures. This one is about what climate change will look like if we don't act. The next one argues that another world is possible.

We don't need a crystal ball to see what climate change will look like. We just need to look at what happens in climate disasters now. The future will be like that, only much worse. This chapter will look at Hurricane Katrina in New Orleans and the long drought in Darfur. My main point is that there are no simple climate disasters. They always take place in the midst of a capitalist system, and that system turns a natural disaster into a human tragedy.[1]

But there is another reason to look at these disasters. When people see what global warming *does*, it can be the springboard for organising to stop global warming. Or it can be the moment when the rich and powerful make the poor pay for the crisis, and the poor turn on each other to fight for what little is left. So I will look at what happened in New Orleans and Darfur with an eye to the political responses of people on the ground, and the sort of responses a climate justice movement could mount in future disasters.

New Orleans

Hurricane Katrina struck the coasts of Louisiana and Mississippi early on Monday morning, 29 August 2005. 80 percent of the city of New Orleans was flooded, and almost 2,000 people died. Far more have died in other tropical storms. But Katrina shows what global warming can do in the richest countries.[2]

Katrina could have happened without climate change. Hurricanes are measured on a scale from Category 5, the worst, to Category 1, the gentlest. Katrina was only Category 2. What global warming does, though, is to make worse Katrinas more likely in future. This is because heat causes hurricanes.

Severe tropical storms are called hurricanes in the Atlantic and cyclones or typhoons in the Pacific and Indian Ocean.[3] What is critical for hurricanes is the heat over the hottest part of the year in one particular place. Hurricanes form and grow over oceans. The intensity of the storm depends on how warm the ocean is, how deep that warmth goes, and how many days the heat lasts.

The real damage, though, comes from the hurricane "surge". The winds, pushing forward, pile up the water on the surface of the ocean.[4] When that wave nears the land, the resistance of shallow water makes the water slow down and rise up just before it hits the beach. The surge from Katrina was 29 feet tall (8.7 metres) when the eye of the hurricane hit the coast of Mississippi.[5] There was enormous power in that wall of water.

Everyone now knows that global warming raises sea levels. This happens because warm water expands, and because heat melts the great ice sheets in Greenland and Antarctica. The danger, though, is not a gradual rise in the waters. It is the moment when that rise combines with the surge of an intense storm to suddenly over-whelm the land's defences. The heat both raises the water and increases the storm.

This had already begun to happen. The sea level off Louisiana had already risen by a foot (0.3 metres). This is trivial compared to what the future holds. But several things had happened on the coast and in New Orleans to amplify the effect of the sea level rise, so that Katrina mimicked a storm driven by much higher seas.

First, the coast and the city of New Orleans sank by 3 feet (0.9 metres) in the 20th century. This was because the city had been pro-tected by a system of levees (dykes). For thousands of years the delta of the Mississippi river had brought down millions of tons of silt that replenished the wetlands between the city and the coast. The levees trapped that silt and directed it straight out in the ocean. Without the silt, as the city and wetlands dried out, the land com-pacted and sank.

Moreover, the wetlands were eroding. Oil and gas are the main industries in southern Louisiana, and the oil companies had honey-combed the wetlands with thousands of channels and canals that ate away at the porous land. By 2005 the wetlands were disappearing at a rate equivalent to three football fields a day. The wetlands had absorbed much of the surge in previous hurricanes. By 2005 much of that protection was gone. The surge from Katrina was 29 feet high when it hit the coast. It was still 14 to 18 feet high when it hit the city's levees 40 miles to the north.

Those levees themselves had been eroded by neoliberalism. The Army Corps of Engineers had always been responsible for the levees.

Alternative Futures

The corps had once been the most respected institution in American engineering. Hundreds of officers had designed and supervised the levees, and tens of thousands of enlisted men had built the flood defences. But with the coming of neoliberalism the corps was hollowed out by subcontracting. In 2005 there were only three army officers left in the corps in Louisiana. The rest of more than 300 central staff were civilians and contractors. All the physical work was done by subcontractors.

Subcontracting public works leads to corruption everywhere in the world, and so it was in Louisiana. This is partly simple bribery. But the real profit in any public contract is to be made by not doing the work to standard. So it was in New Orleans too. The contractors built and maintained the levees all right. But the steel structures inside the dykes were of cheaper kinds that could not withstand the surge. The corps had traditionally made sure that the earth under the levees was built up and reinforced to hold the steel. The contractors built on whatever porous clay or mush they found. In the words of two New Orleans journalists, it was like "putting bricks on jell-o".[6] When the surge from Katrina hit, the levees broke in hundreds of places. They were effectively no longer there.

The result was that Katrina had the same effect as a hurricane with a far larger sea level rise. The sea level had risen by a foot. The land had sunk by 3 feet. That made for an effective rise of 4 feet, and 80 percent of New Orleans behind the levees was below sea level. The levees which should have worked were 14 feet high. That meant an effective sea level rise of 18 feet. And the erosion of the wetlands amplified even that. So Katrina had the same effect as a hurricane would with 20 to 25 times the rise in sea levels we have already seen. That is why it gives a picture of the future.[7]

This was not an unexpected tragedy. Everyone in Louisiana knew what was coming. They didn't know the levees were rotten and they mostly didn't understand about climate change. But they knew the land was sinking and the seas were rising. The local paper, the *New York Times*, many other newspapers, CBS and PBS all ran stories on it.[8]

Everyone had the same name for what was coming—the "Big One". The Federal Emergency Management Agency (FEMA) in Washington was the part of the government tasked with responding to disasters. In 2001, before 9/11 and Katrina, a FEMA report said the three biggest disasters threatening the US were a terrorist strike in New York, an earthquake in San Francisco and a hurricane in New Orleans.[9]

In July 2004 FEMA sponsored a week long war game exercise in Louisiana on a virtual "Hurricane Pam". FEMA director Michael Brown opened the proceedings. Two hundred and seventy hurricane scientists, Coast Guard, national, state, city, and town officials attended. Their computer model hurricane, "Pam", was a Category 3. In the war game "Pam" flooded the whole city of New Orleans and killed 60,000 virtual people across the region.

Everyone knew what had to be done too. Scientists at Louisiana State University had drawn up the plans for flood defences and a network of canals that would replenish the silt in the wetlands. The estimated price was $14 billion—the cost of six weeks of the Iraq War. Republican and Democratic governors of Louisiana had begged the federal government to do the work. Congress, President Clinton and President Bush had refused to spend the money. The Bush administration had even cut the flood defence budget of the corps to pay for the Iraq War.

Then, on top of everything else, Mayor Nagin and Governor Blanco failed to evacuate tens of thousands of New Orleans residents in the path of the storm. On Saturday, two days before Katrina hit, it still looked like a Category 5 hurricane—the "Big One". But Mayor Nagin was under pressure from the hotel industry. New Orleans was a tourist city, and the hotels had room for 30,000 guests. If the mayor ordered a complete evacuation, all the tourists would leave and the hotels would lose millions.

Nagin was an African-American and a Democrat. He was also the business candidate for mayor. On Saturday he bowed to the hotel industry and issued a "voluntary evacuation" order—leave if you want to.

Max Mayfield, the director of the National Hurricane Centre, rang everyone he could to get them to persuade the mayor to act. Mayfield got through to Mike Brown, the director of FEMA. Very worried, Brown called President Bush at his Texas ranch early on Sunday morning. Bush called Mayor Nagin. Under presidential pressure Nagin backed down and ordered a "mandatory evacuation" on Sunday morning.

Later Bush would try to conceal that phone call. We will see why.

In any case, Nagin's "mandatory evacuation" still really meant, "Leave if you want to." The mayor told people to drive out of town. But a quarter of New Orleans households had no cars. The city had a fleet of 550 city buses and 264 school buses with a total of 48,000 seats.[10] The buses were left on the lot. The commercial greyhound buses and the last train out of town left empty to protect

Alternative Futures

the investment. Many people with cars were reluctant to leave too. There were no shelters set up to shelter and feed them. It was the end of the month, and people were waiting for welfare cheques, unemployment checks and social security pension cheques. So the reckless, the old, the disabled and the poor tended to stay put.

It didn't have to be that way. Half the people who died in New Orleans drowned in the Lower Ninth Ward. The director of the animal shelter in the Lower Ninth Ward, Laura Maloney, made sure that every one of the 263 stray dogs, cats and pets in her care was evacuated safely to a shelter in Houston.[11] The sheriff of Lafayette Parish, closer to the coast, made sure his deputies went house to house urging people to leave, and provided the buses to take them away. But if Nagin had done that, it would have emptied the hotels on the high ground in the city centre.

When the hurricane hit and the levees broke, hundreds drowned within the hour. Many more had axes, and hacked through the attic ceiling and climbed out on the roof to wait for help. And waited. And waited.

Almost everything that should have happened then didn't happen. The reason was what a generation of neoliberal policies had done to all American cities, not just New Orleans. For 20 years city and state governments had cut welfare programmes and all forms of aid for the poor, the elderly and the disabled. Instead mayors saw their job as saving money and bringing affluent residences and businesses back into the central city. Senior managers had long since stopped listening when their workers on the ground complained about what they couldn't provide. To ram through neoliberal policies, managers had to ignore the cops on the beat, the emergency technicians, the classroom teachers and case workers. When disasters struck, the managers were already hard wired to ignore the frantic phone calls from their people on the streets.

Standard operating practice with all the suffering caused by neoliberal cuts and neglect was to ignore it. If the press finally exposed it, the managers would have a crisis meeting. That meeting was not about how to reverse the policies and help people. It was about what to tell the public. The first line of defence was always to lie about what was happening. The second line of defence was to blame the victims. What they did for all neoliberal cuts was what they naturally did in disasters. It's what the city government did in the heat wave that killed more than 700 elderly people in Chicago in 1995. It's what the federal government and Donald Rumsfeld did when Baghdad fell in 2003. And it's what happened in New Orleans.[12]

The police and fire service were told not to help the stranded. The Coast Guard still rescue people daily, so they did it in New Orleans. Thousands of ordinary people put their small boats into the water and went looking for people to rescue. The Federal Emergency Management Authority forbade them to do it. The folk in the small boats ignored FEMA. They took everyone they rescued to the city centre, or left them on highway flyovers. There they joined survivors who had waded and swum through the water to high ground. Where they waited again, for days.

FEMA turned back doctors, firefighters and emergency personnel who came from round the country to help. The navy sent a hospital ship, the USS *Bataan*, to stand off the coast, and President Bush refused to give it permission to evacuate the sick and wounded. The major bus companies offered to help, and FEMA did not return their calls.[13]

It took almost 12 hours for the waters to rise in most of New Orleans. During that time the city and federal government denied that the levees had broken. If they had told the truth, hundreds of lives would have been saved. FEMA had only one man on the ground, Ron Brown, but he was using a satellite phone to tell Washington what was happening. They ignored him.

By the next day, Tuesday, Bush and his people were getting their act in gear. That didn't mean helping anyone. It meant minimising the problem, telling the press FEMA was doing "a heck of a job", and staging disastrous photo ops. Bush and Brown pretended they didn't know what was going on. They did. Brown had rehearsed "Hurricane Pam" and its virtual 60,000 dead. He had called Bush in a panic the morning before Katrina, and briefed Bush by video call that same lunchtime. But Bush would do everything he could to keep those phone calls and briefings secret for months afterwards.

On the face of it, this doesn't make sense. It made Bush look stupid. By Tuesday the world's press were in New Orleans showing the world what was happening, and Bush and Brown still said they didn't know. But their main concern was to conceal that they knew beforehand what would happen, had done nothing for years and had not evacuated people. And Bush had a special problem. He was the world's leading climate change denier, and the leader of opposition to action on global warming. The media were denying Katrina had anything to do with climate change. But an awful lot of people were thinking, if that's not climate change, what will the real thing be like? So Bush had to minimise what was happening.

That didn't work. The next line of defence was blaming the victim. If you looked at the television pictures, the tens of thousands of exhausted and homeless survivors in New Orleans were mostly African-Americans. The day after the storm the American newspapers and television went into racist hyper-drive. They said the survivors were an army of looters, an angry mob. Snipers were shooting at rescuers. The police chief of New Orleans, an African-American himself, went on Oprah Winfrey and said there were men running around raping children.

The rape, the snipers, the violence and the mobs were lies. The looting was real enough. People needed food, water and nappies, and the stores were closed and no help was coming. Many people, especially cops, were taking more valuable things as well. But the message of the media was these people have caused their own suffering. They don't need help. They need control.

The reporters were pouring into New Orleans along open roads. When a black crowd tried to march out of the city across the bridge into a mainly white suburb, they were met by a wall of armed police. The cops shouted "Nigger!" and fired over their heads. When people sat down on the bridge, the cops held guns to their faces and moved them back.

For two days racism monopolised the American media. And it didn't work. The reporters on the ground grew outraged, and the American public saw through it. Faced with a choice between the President of the United States and crowds of desperate poor black people, most white Americans didn't side with the white guy.

By Thursday television anchors were savaging the government. Mayor Nagin broke ranks that night. He called a local radio show, and said he was pissed that no help had arrived. He said the Iraqis didn't have to beg for American troops. On Friday night the rapper Kanye West appeared on a national telethon. West departed from his script and talked about race, poverty and the Iraq War, and then shouted, "George Bush doesn't care about black people." By Friday morning the buses finally arrived and began evacuating survivors.

More than that, this was the week George Bush lost America. Public opinion had already been turning against the Iraq War. After 9/11 Bush had flown to New York and cast himself as the spokesman of the first responders and the American people, vowing revenge. New Orleans was a disaster on the same scale. Bush wouldn't go talk to the people there. His handlers were afraid of what angry people would say in front of the cameras. And Americans could see the government wasn't helping. It wasn't just black people Bush didn't care about—he didn't care about ordinary Americans. His public standing would never recover.

Hurricane Katrina and the government non-response also transformed the public debate in the US on climate change. During that week the media mentioned climate change only to say it had nothing to do with Katrina. But in the months afterwards papers began appearing in scientific journals showing that global warming had doubled the intensity of tropical storms. A few months after Katrina the eminent hurricane scientist Kerry Emmanuel, speaking to an audience of 800 scientists, said it was time for government scientists to tell the truth about climate change. The audience cheered.[14] Outside the scientific community, New Orleans made many people think seriously about what climate change would mean. The way was open for Al Gore's impassioned film. The public, and politicians, began saying that something must be done. Bush grew increasingly isolated, and finally had to admit that, yes, it is getting warmer.

Race and reconstruction

But the tragedy of New Orleans did not end with the evacuation. There was no help for people to return and almost no temporary housing. Two years later half of the people had not come back. Those who did return found there was almost no help with reconstruction. The city had little money and almost no federal support. A large proportion of city employees were fired. It was many months before the debris and garbage were cleared, and the city still looks devastated. The government kept promising to restore the levees. But by 2008 they had not even been restored enough to withstand a Category 3 hurricane. The government had no plans to build them strong enough for a Category 5, which everyone knew was coming sooner or later. And there were no plans for the expensive flood defences and network of channels needed to restore the wetlands and stop the city sinking.

New Orleans had been abandoned. The pervasive feeling in the city was depression. People lived on pills and cried in supermarkets. Suicide was common. Everyone had a story to tell, and the details were always different, but the meaning was always the same—no one cares about us.[15]

Most people, in New Orleans and nationally, thought the government was ignoring the city because most residents were black. Most working class African-Americans in the city thought their

Alternative Futures

neighbourhoods were being left to rot because the real estate developers wanted to drive poor blacks out and replace them with upmarket housing for white professionals. This was a natural thing to think—the developers had long wanted to do just that. And racism is real. But to look only at racism is to mistake the scale of the problem. The federal government knew perfectly well that they had no real plans to stop global warming. Given that, the seas were going to rise and the hurricanes increase in intensity. To save the city, they would have to build levees and flood defences to withstand bigger storms than anyone had ever seen. And given that, they had in effect decided to abandon the city.

Climate campaigners had long thought that the first places lost to global warming would be atoll nations in the Pacific and Indian Ocean. We were wrong. New Orleans was the first place to go.

Most people in New Orleans, the majority of them from the black working class, could not get the measure of what had been done to them. This was because they thought largely in terms of race. So did the left, the liberals and the African-American intellectuals nationally. This confused them. What had happened was that racism had been used to divide black and white victims and survivors.[16]

It's worth looking at the pattern of death in the hurricane. The majority of the dead were black. But that fact hid the white people who died, mainly in the suburbs and along the coast of Mississippi and Louisiana. Among both blacks and whites the poor were far more likely to die, and the elderly poor were most at risk. Comfortable middle class and professional African-Americans got out and did not find themselves marooned in the stadium or on the bypasses. It was class that killed.[17]

But it was not only really poor people whose lives were wrecked. What destroyed southern Louisiana was that the insurance companies ruled that the damage came from flooding, not from the storm. Their insurance did not cover storm damage, and only a quarter of home owners had federal flood insurance. So most people, renters and home owners, lost everything. Given the state of the flood defences, new mortgages were unavailable in areas that would be flooded again. It was the insurance companies and the banks who destroyed the region. And the federal government and the courts allowed them to do it. The really affluent could cope with that. The majority, black and white, could not.

I am not saying that racism is not real. But racism has long been used in America to divide people. For instance, racism and the stigmatisation of black single mothers have been used to justify shredding the welfare net in America. It is true that black people are more likely to be poor, unemployed or on welfare. But it is also true that the majority of black people are not poor, and the majority of the poor are white. The majority of the unemployed and welfare recipients are also white. Moreover, the majority of the poor and people on welfare are only in that position for a few months or years, and the majority of white working people find themselves needing government help at some point in their lives. Racism has justified the attack on welfare and government programmes, but that attack has hurt white working people too.

Racism worked the same way with Hurricane Katrina. At first the media justified the government not helping anyone, black or white, because the survivors were a dangerous black mob. After the hurricane the real racism in the city concealed what the government, the banks and the insurance companies were doing to all working people.

It's important to understand clearly because worse hurricanes are coming. The great population concentrations most at risk in the US are in New Orleans, Houston, Miami and New York. New York would see the most devastation, because there are more people, Manhattan is very hard to evacuate, the subway system will flood, the rivers provide a perfect channel for hurricane surges, and many people will be trapped in dangerous high rise buildings. So we need to look at the sort of resistance that was missing in New Orleans.

Part of what was missing can be seen in the crowds of mostly black survivors trapped in the stadium, in the Convention Centre and on the streets of the city centre. Forty years ago, in the time of the civil rights movement, those people would have organised themselves. In 2005 the leaders of the black community, the politicians, professionals and preachers, simply were not there. Working class black people had lost the habit of organising themselves.

But imagine if there had been 12 people there who were organised with each other, in the habit of organising others, and who saw themselves as fighting for social justice and against climate change. They could have started with small meetings in different parts of the stadium. Then they could have turned the stadium into a meeting. They could have elected representatives to speak to the media of the world. They could have said to the cameras, "Soldiers, firefighters, bus drivers, truck drivers of America, drop whatever you're doing, and drive to New Orleans. Rescue us." Those people would have come, in their tens of thousands, and Bush would have raced to get the help there before them.

Alternative Futures

As in the time of civil rights, the activists could have led tens of thousands of people up to the cops who guarded the bridges, and marched through them singing, watched by the cameras of the world. And at the same time, because they understood climate change, they could have mobilised the sympathy of most Americans in that moment to demand the money be spent on new levees, flood defences and rebuilding. And they could have said none of that will matter unless the world does something about climate change.

That may sound fanciful, because it didn't happen. And it is no criticism of the survivors in New Orleans that it did not. They were stuck with the limits of opposition politics in America in 2005. But it is not fanciful if climate activists and social justice activists start building networks now in the knowledge that a time will come in the future when we will have to intervene in that way. This is true of other countries just as much as it is in the US. There will be great climate tragedies. But they will also be moments when a grassroots movement can change history. The key will be networks and organisations that fight for both social justice and climate justice. They will have to be daring, and believe it is possible to mobilise large numbers of people and appeal to the world. They will also have to organise simultaneously to defend the survivors locally and to fight climate change globally.

Darfur

New Orleans is an example of what climate change can do in a rich country. Darfur shows what can happen in a poor country. Again, my central point is that there is no such thing as a simple climate change disaster. What has happened in Darfur was a climate tragedy. It was also a consequence of neoliberalism and global competition for Sudanese oil reserves.

I'll start with the economy and oil, and then go on to climate change. Darfur is the westernmost province of Sudan. At independence in 1956 the British colonial authorities left Sudan a poor and underdeveloped country with only two factories—a brick works and a beer brewery. Sudanese politics since independence have been a shifting kaleidoscope of elected governments and military dictatorships. But all these governments have been led by people from the same small elite. And there has been a basic continuity in the policies of the governments in the capital, Khartoum. They have tried

with increasing desperation to administer a poor and underdeveloped part of a global system in economic trouble.[18]

After independence the government in Khartoum borrowed heavily from foreign banks and governments to develop large farms growing cotton for export along the Nile Valley. That left the rest of the county poor. Then in the 1970s the world economy went into the long crisis because of the fall in industrial profits. The Sudanese government was suddenly unable to pay back the international loans. The price of cotton fell, so export earnings fell too, and the interest rates charged by global banks rose.

The answer from the governments of the West was bailout loans from the International Monetary Fund (the IMF) in the

Alternative Futures

1970s and 1980s. But, as everywhere else, the IMF demanded strict conditions for the loans. All over Africa the fund insisted that governments cut their expenditure, reduce public services, hold down wages, and cut subsidies for basic foods, fuel and cooking oil. They also insisted on deregulation of markets and foreign exchange, contracting out government programmes, and charging fees for education and healthcare, in other words, the whole neoliberal package.

The result of all these measures was to impoverish the majority of Sudanese. For instance, the real value of the minimum wage in 1986 was only 17 percent of what it had been in 1970.[19] But in a wrecked economy the central government had even more trouble raising tax revenues.

Then in 1984 the government in Khartoum saw the solution to all their troubles. Oil had been discovered in southern Sudan. The central government had to get their hands on it. That meant civil war. To explain why, we need a bit of history and geography.

Sudan can be roughly divided into North and South. The North contains the valley of the Nile, the capital, Khartoum, and most of the fertile irrigated land. Wealth is concentrated in the Nile Valley, and almost all the elite who have run Sudan under various governments came from here. Some northerners are Arabic speakers and some speak other African languages, but almost all of them are Muslim. Darfur is the westernmost, and the poorest, of the Northern provinces.

The South is the poorest part of the country. Under the British the South had been kept separate, economically backward and uneducated. The people practised traditional African religions or Christianity, not Islam, and few spoke Arabic.

At independence there was a long civil war between the northern government and southern rebels. This was settled by a peace agreement between North and South in 1972. That peace agreement held for 12 years—war between North and South was not inevitable. But the peace agreement gave the southern regional government control of any future mineral revenues. That didn't matter then—no one knew the South had oil. By 1984 it mattered a great deal. The oil was there, and it would have to be piped out through the North. The impoverished central government in Khartoum had to destroy the southern regional government to get control of the oil fields. The civil war between North and South was on again. The main oil company was Chevron, an American multinational, and so the US backed the North.

Climate change and drought

By 1984 the country was impoverished, the government broke, and a bitter civil war was beginning for control of the oil fields. This political situation changed a climate disaster in Darfur into a terrible famine and a long war.

Darfur is home to three million people. It's in the Sahel, the belt of semi-arid land across Africa that lies south of the Sahara. In the late 1960s the rains failed in the Sahel and Darfur, and they have never recovered. The reason was climate change. Traditionally the rains came because the Atlantic Ocean off West Africa was warmer than the Indian Ocean off East Africa. The difference in temperature drew wet winds from west to east, and as they crossed the Sahel and Darfur they dropped their rain. Then climate change warmed the Indian Ocean.[20] There was no longer a temperature difference to pull the winds, and the rains became more erratic. There is less rainfall, it falls more often outside the rainy season, and it is more likely to fall in bursts of downpours that run off the land, rather than steady rains that nourish crops and grass. In the good years people live closer to the edge than they once did. In the bad years famine threatens.

This is part of a larger pattern that we can expect with global warming, and already see beginning to emerge. As the world grows hotter, the bands of good rainfall in the temperate countries will move towards the poles. We have already seen heat waves and droughts moving into the southern Mediterranean, Australia and the south western United States. The result is likely to be that temperate Northern Europe gets more rainfall, and many of the poor countries less.

In Darfur and the neighbouring regions the changes in rainfall also had a very particular effect. Traditionally, some Darfuri groups were mainly herders of camels, cattle, sheep and goats. They migrated each year from dry season pastures to wet season ones, following the grass. Other Darfuri villagers stayed put and farmed crops fed by rain. The climate change drought has now lasted 40 years. In a very bad year the crops now fail. In a good year the farmers have enough rain. But the seasonal grasses the herders depend upon are more vulnerable to the drought. Even in a good year the herders have trouble finding enough grass. In a bad year they have trouble even finding enough water to keep their animals alive.

If the crops fail, some villagers die. But when the rains return, the villagers still have their land and can farm. Herding households are in danger of losing all their animals and being reduced to permanent

Alternative Futures

destitution. The only solution is for the herders to use the pastures and fields of the villagers and take the water from their wells. That can only be done by violence. The current war in Darfur is in part the culmination of 40 years of herders and farmers fighting each other for grass.

Traditionally, Darfur had been home to a patchwork quilt of ethnic groups.[21] In the North there were camel herding Arab speaking tribes, in the South cattle herding "Baggara" Arab tribes. In the centre and south the largest group were farmers who spoke dialects of the "Fur" language. But the picture was much more complicated than that. People were connected to each other by all sorts of ties. "Fur" would often take up herding and become "Arabs". "Arabs" who lost their herds would become "Fur". There were also Fur herders, Arab farmers and Arab traders. Everyone was a Muslim, everyone was dark skinned, and most people spoke some Arabic. Most Fur and Arabs supported the same moderate Islamist political party. Besides the "Fur" there were many other smaller "African" groups of herders and farmers. And in each region a mosaic of peoples lived side by side.

Moreover, the "Arab" herders had reasonably friendly relations with the villages of Fur and other farmers along their annual migration routes. The Arabs would look after animals for the farmers, graze their herds on the stubble after the harvest, and exchange presents and payments for water and grazing rights. There were occasional fights with spears and arrows, but these were settled by the government bringing together clan elders from both sides.

All that changed in the famine of 1984-85.[22] For a generation the drought had been eroding the pastures of the camel herders in northern Darfur, along the edges of the Sahara. But the full impact of the drought only hit the farmers and cattle herders of southern Darfur from 1983. By 1984 the crops had failed two years running, everyone was hungry and the animals were dying.

In 1985 famine relief was no longer the business of foreign governments. It would be done by foreign aid and foreign NGOs. There was nothing particularly Sudanese about this. The IMF, the US and the European powers were insisting on it everywhere. In Sudan the American government aid agency, US AID, was supplying 80 percent of food aid. Gerard Prunier, the historian of modern Darfur, takes up the story:

> The Americans had decided to give a quasi-monopoly for transporting the food to a strange joint venture between a businessman from

Baton Rouge, Louisiana, and a Lebanese wholesale grocer in Khartoum. The resulting company, Arkell-Talab, had proved remarkably inefficient and out of the 68,000 tons of food aid allocated monthly to Kordofan and Darfur only 123,000 tons had been transported between December 1984 and June 1985, ie an average of a little more than 20,000 tons a month, barely a third of the planned figure. Most of the 20,000 trucks which were supposed to be used for food convoys were in fact out of commission, and people continued to starve in the midst of plenty... [Afterwards] everybody knew that [the famine] had been entirely preventable.[23]

About 100,000 thousand people died. Most did not directly starve, but died of disease instead. The reason was that both farmers and herders tried to hang onto what they had. Most farmers had rights to land as part of a village community and were terrified they would lose their land permanently if they left. Most herders also stuck to the old migration routes and tried to save some of their animals. Both groups were afraid of permanent destitution, and as little food was being distributed there was little reason to go the towns. Moreover, most areas had supplies of wild food, particularly berries, that people could gather and have some nourishment.

The result was not starvation in most areas, but widespread malnutrition. This opened the way to epidemic diseases. The drought also meant that much of the water was dirty. Disease carried off the weakest, mostly old people and small children.[24]

Many pastures simply dried up. Migrating herders had to take their flocks on new routes to new areas, where they had no traditional arrangements. In many cases the herders' harvest from their own small farms at home had failed. The animals were now all they had. The farmers needed what pasture remained for their own animals. In extreme cases, the migrating herders wanted to use the fields of the farmers to pasture their flocks and took them by force.

There was nothing traditional about these conflicts. This was life and death. It was also a question of whose family would face permanent want.[25]

By the autumn of 1985 the famine was over in Darfur. The rains had come and the harvest was good. But people were left with their feelings. People make hard choices in famines, and say no to people they have known all their lives. These are the refusers. Sometimes they deny food to their own parents, or take food ahead of their children. In many famines more girls die than boys. In Darfur, where

Alternative Futures

gender relations were relatively equal and women and girls controlled the cooking, more boys died.

Then there are the refused. People were turned away from food and water, and were helpless to stop it happening. All this will have left people with guilt and hatred.[26] Imagine the following. In 1985 a farmer denies a herder water and pasture. Most of the herder's animals die, and then two of his three children die of disease. The next year the herder comes again to that same village, with his few remaining animals. The same farmer stands in front of him, barring the way to pasture. The herder recognises the man's face. He also remembers the face of the dead children he could not save. And this time he has a machine gun.

After 1985 the herders knew they had to do whatever was necessary to stake out new pastures. In the two years after the famine, 1986 and 1987, the rains were good enough. But the fights over migration routes and pasture were worse than in the famine year. Moreover, the herders now had automatic weapons, and they had a new form of racism to justify attacking the farmers. The weapons and the racism were coming from two directions. Both were a result of the politics of oil.

Guns, Chevron and Libya

The first source of guns was the growing war between the northern government in Khartoum and the southern rebels. The prize was the oil fields. Most of the oil was just inside southern Sudan, close to Darfur and the neighbouring province of Kordofan. Chevron, an American company and one of the largest multinationals on earth, had the contract there.

The southern rebels held the land south of the oil fields. No one held the oil fields but the "African" farmers and herders who lived there. The Sudanese government, though, was broke. It could not pay or arm enough soldiers to take over the oil fields. So late in 1985 the Sudanese army and the Chevron oil company went to the Arab herders of Kordofan and Darfur who had lost most of their herds in the famine. The army and Chevron offered the herders a deal. The herders would join an unofficial militia. The army and Chevron would give them money and machine guns. They would attack the "Africans" who were living in the oil fields, destroy their villages and crops, and drive them out permanently. The herders would take their animals and have permanent control of the pastures in the oil fields.[27]

The traditional clan leaders, the richest men among the herders, were mostly opposed to the deal. But the young men with nothing joined the militias. What they did in the oil fields brutalised them.

In December 1988 Ushari Mahmud, a linguist and human rights campaigner from northern Sudan, explained to the British researcher David Keen the economic importance of the raids by Rizeigat camel herders from Darfur in the oil fields:

> The Rizeigat militia leadership would go to one [Rizeigat] village and start shooting in the air for two or three days. Rizeigat men would come there from different villages, on camels, on donkeys, on horses, with different kinds of guns. Then they would get their food supplies. Then they would go toward the Dinka [local villagers] area. They spread around a village very early in the morning and start burning around the village, burning the crops, and then they attack. The pattern is that men run away. Hardly anyone puts up a fight. They get the wives and the children. They burn everything. They cut the fruit trees down. They want to make sure that nobody comes back to live in the area. Then they come back, and north of Safaha the distribution [of cattle and people] starts. This is done strictly according to your firepower and hierarchy and if you had a horse or donkey, your role in combat, how many people you captured—a very complex system of who gets what—women, children, and cattle. Only children who are being breast-fed are allowed to go with their mothers. You always find families dispersed.[28]

After the raids came the famines. The oil fields were a wasteland. The younger men walked hundreds of miles across war zones to Ethiopia. The women, children and old men hid out in the bush. Months later, desperately hungry, they began to walk north towards refugee camps. There the Sudanese army and the militia trapped them, and tens of thousands starved to death.

For the moment the oil fields were cleansed. But the men from the villages returned as armed rebels. Chevron lost control of the oil fields, and the militias of Arab herders from Darfur could not use the pastures. So the men from Darfur returned home in the three years after the famine. But now they had machine guns, a tradition of organised looting and a new ideology of racism. The political justification for cleansing the oil fields had been that the people there were black "Africans", and the herders were superior "Arabs" who deserved to rule.

The guns, the organisation and the racism were now turned against the Fur village farmers in Darfur. If the young men with nothing couldn't get pastures to the South, they could seize them in Darfur.

That was one way the arms and the racism came to Darfur. But there was another way too—from the war in Chad.[29]

I am aware that my explanation of the roots of the Darfur tragedy may by now seem quite complicated. But bear with me. The complication is the point. On one level this was a simple climate change tragedy. On another level, great and small powers from all over the world were fighting proxy wars over oil in and around Darfur. It was the intersection of the local war for grass and the global war for oil that created horror in Darfur. It's important to grasp this because future climate change tragedies will be the same. The disasters will happen in the midst of all the divisions and competitions that already afflict so much of the world.

Chad is the country to the west of Darfur. There was a civil war in Chad from 1968 on that kept spilling over the border into Darfur. The border was just a line on the map. The tribes and ethnic groups on both sides were the same. People had always moved back and forth, migrating with their animals or looking for new land. When the war in Chad was at its worst, armed refugees spilled over into Darfur. And when Darfur was bloodiest, armed refugees fled to Chad.

The war in Chad has now lasted most of the last 40 years. It is partly another climate change war. Chad has endured the same long drought. Northern Chad is dominated by Arab and "African" tribes of herders. Southern Chad is dominated by "African" farmers. Both groups have been fighting for land, water, pasture and grass.

But from 1968 to 2004 the war in Chad was also a proxy war between Libya on one side and the United States and France on the other. Libya's leader, Colonel Gaddafi, was also trying to extend his power. In Chad he backed the northern herders with arms and soldiers in the name of Arab nationalism. France, the old colonial power in Chad, backed the southern "African" villagers. Israel sent special forces and aid, on the grounds that the Africans were not Arabs. And from the late 1970s on, the United States became the main backer of southern Chad governments as a way to make trouble for Libya.

Libyan army units and armed herders from Chad regularly took refuge in Darfur. At the worst point of the famine in 1985, when the Americans were sending no food aid, the Libyan army arrived with in Darfur with large shipments of grain. The Libyan army also allied themselves with the Arab herders in Darfur. Like Chevron, they gave the herders machine guns. Like the Sudanese government, the Libyan army told the herders they were Arabs, and it was their historic destiny to fight the African farmers.

As poor herders armed and organised to take the pastures from the farmers, the largest groups of farmers, the Fur, got arms from the government of southern Chad and dug in to defend themselves. By 1988 war was spreading across central Darfur. Darfur had once been a complex patchwork of ethnic groups. Climate change and international intervention had simplified that into a war between Arabs and Africans. In Prunier's phrase, Darfur had seen "the globalisation of ethnic cleansing".[30]

At this point the Sudanese government finally intervened to broker negotiations in Darfur. The traditional chiefs of the various tribes came together and made peace in 1989. It was what most people in Darfur—Arab and Fur, militia and civilians—wanted. That peace would hold for 14 years. It is the single most important piece of evidence that left to themselves the people of Darfur can live with each other.

But the memory of war remained too. The militias kept most of their arms. The Fur arms went underground—literally, they buried them. There was another famine in the early 1990s. There was little aid, no international interest, and noone counted how many died. The long drought continued year after year. The squeeze on pasture and farming land must have ripped away at people's nerves and livelihoods.

And no one would leave Darfur alone.

Globalised war

This brings us to the current war.[31] The new war in Darfur is usually portrayed as a genocide of Arabs against Africans, motivated by racism and Islamic fundamentalism. The reality is different, more complex and tragic.

Climate change is still at the centre of the story. The drought has continued. The herders of northern Darfur have lost more of their flocks. Many have migrated south to find pastures or work. In southern Darfur both the farmers and the Arab cattle herders have also found their livelihoods under increasing pressure. Everywhere increasing numbers of people without either land or flocks are look-

Alternative Futures

ing for established landholding villagers to allow them access to a living. But the villagers no longer have enough to share.

Across Darfur there are also increasing numbers of refugees from war and climate change in Chad. These people are destitute, heavily armed and supported by the Libyan army.

All this means that relatively low level conflict over land and grazing has been chronic. Darfuris have struggled as best they can to contain this conflict.

The continuing grip of neoliberalism means that there is no economic security. Darfur has grown even poorer. The landless cannot find jobs in the towns. And people know they cannot look to the government for famine relief in the bad years. There is no help for people facing climate change.

There is also political insecurity. The Sudanese government is still desperately strapped for money, because neoliberal globalisation is impoverishing Sudan and much of Africa. The government has also been embroiled in an expensive war in the South. Once tribes caught up in conflicts over land and pasture could look to the central government. The police would arrive to contain fighting. The centrally appointed governor of Darfur would then mediate between the warring groups, as happened in 1989. Now the central government lacks the money, troops and will to play that role. Darfuris find themselves left without any security to fall back on.

Then there is international meddling. From 1991 to 2000 the military dictatorship in Sudan was led by Islamists. They were enemies of the United States. Chevron, seeing no hope, left Sudan in the early 1990s. Then in 2000 the Islamists in the military dictatorship split. The dominant faction of army officers decided to make peace with the southern rebels and become allies of the United States. From 2001 on, Sudan became an ally of the US in the War on Terror. A peace agreement with the South provided for a temporary federal solution. Crucially, the oil would be moved by pipelines across northern Sudan to the Red Sea. The government in Khartoum would get half the oil revenue, and the new southern government of the rebels the other half. Libya also made its peace with the US, and Gaddafi became an American ally.

Most of the new oil contracts in Sudan went to Chinese state owned oil companies. China was building alliances across Africa to control resources essential to a growing Chinese economy. So China became the main ally, and financial support, of the Sudanese government.[32]

The consequences of these shifting alliances created tragedy in Darfur from 2000 on. When the military dictatorship in Sudan began to align itself with the US from 2000, the more "hard line" Islamists in Khartoum lost much of their influence. Their leader, Hassan Al-Turabi, turned to "Africans" in Darfur for support. Turabi and his followers in Darfur began arguing that the permanent problem in Sudan was that it was controlled by an elite based on tribes around Khartoum. This elite, they said, had always controlled the central government and taken most of the wealth for their own area. Darfur had been left out and impoverished. So people in Darfur should rise against the government to get their share (and support Al-Turabi).

This analysis resonated in Darfur. An alliance grew between "African" villagers in Darfur trying to hold onto their land and local Islamist intellectuals. The Fur militias began to dig up their buried arms. They also got more arms. It is unclear where the money was coming from.[33] But the key thing was that many villagers in Darfur felt their backs were to the wall.

In the spring of 2003 the Fur militias went on the offensive against the central government. Armed with "technicals"—Toyota Land Cruisers with machine guns mounted on the roof—they attacked Sudanese army garrisons with stunning success, killing more than a thousand soldiers.[34]

The Sudanese government felt they had to crush the uprising of the farmers. But they were still strapped for cash, and could not pay or arm enough soldiers. Moreover, many of their soldiers came from Darfur and could not be relied upon. So the Sudanese government turned again to the old militias of Arab herders, now called the "Janjaweed". The core of the Janjaweed were poor men from camel herding tribes in Northern Darfur. The Sudanese army promised them the land and pastures of the Fur farmers.

The Sudanese army strategy was standard counter-insurgency— clear the rural areas of the people who could provide a base for the opposition. They began to drive out the farmers. Typically, the Sudanese armed forces would send in old Russian Antonov cargo planes that dropped burning oil drums on a village—the army could not afford bombs. Then the Janjaweed militia would ride into the village on horses and pickup trucks. They would kill the men, rape and kill the women, and destroy the trees and crops so no one could return. The Janjaweed shouted at the villagers that the government was backing them and now the land was theirs. The villagers fled to refugee camps in Darfur and across the border in Chad. Roughly 200,000 to 400,000 people have died from war, hunger and disease in the refugee camps.

Alternative Futures

The United States government then intervened and convened peace negotiations. Elite circles in Washington were divided about which side to support. The Pentagon, the State Department and George Bush wanted to preserve the alliance with the military regime in Khartoum. For geographical reasons, Sudan was a key player in the War on Terror. But Washington was also competing with China for the control of Sudanese oil, and control of oil and gas right across Africa.

Moreover, several groups in the US were backing the rebels and wanted the US to intervene militarily. Christian evangelist Republicans had long backed the southern rebels against the Muslim north, and in the same spirit continued to back the Darfur rebels. The Black Caucus in Congress backed the rebels on the grounds that they were Africans. And Zionist supporters of Israel loudly accused the Sudanese government and the Janjaweed of genocide.[35]

This combination of the Black Caucus, evangelical Christians and friends of Israel was able to build a good deal of media support in the US for intervention. But power in the United States does not lie with any of those groups. The Pentagon, the oil companies, the State Department and the Bush administration did not want another American war on top of the Iraq debacle. The Bush administration negotiated a peace deal in 2006 which left the Sudanese government in effective control. Most of the Darfuri rebels rejected it, and Washington washed its hands of Sudan.

On the ground, in Darfur, the war and the refugee camps remained. Most Darfuris, Arab and African, longed for peace, and almost all of them came to blame the Sudanese government for setting them against each other. But on their own they have not been able to make peace. And so for now Darfur remains a tragedy and a horror.

Alternatives

Climate change had produced a long drought that set Darfuris against each other for land, water and grass. Neoliberalism and the impoverishment of Sudan then turned that drought into famine. Outside powers intervened to encourage desperate Darfuris to make war upon each other. Once people had lived together. Now globalisation had encouraged ethnic cleansing. It is an awful warning of what climate change can easily lead to.

Moreover, this was ordinary, slow climate change. Abrupt climate change, when it comes, will affect many parts of the world at the same time in many different ways. Crucially, the changes in rainfall patterns will drive up the global price of grain. When that happens, people will starve in many places, and epidemics will sweep through weakened populations. This will create waves of refugees. But this chaos will be afflicting many places, swiftly and unexpectedly, and will combine with storms, floods and heat waves. Sudden mass desperation and government failures will lead to war. Peoples and governments will fall out along previously existing fault lines.

Moreover, we can expect most governments and elites to act like the governments of the Sudan and the United States. They will encourage ordinary people to blame each other for their tragedies, and to fight each other for what little is left in the world's shrinking wealth. It will not be easy for everyone simply to say that the problem is global warming.

It could, however, have been different in Sudan, and there will be alternatives even in the midst of abrupt climate change. Those alternatives, however, will require a mass movement that understands both climate change and social justice.

The people in Darfur know about climate change. In 1985 a young anthropologist from Oxford, Alex De Waal, went to Darfur to study the famine there. The farmers and herders all told him the same thing. The climate had changed, and that was the root cause of the famine. The rains did not come, and when they did they fell at the wrong time of year and in tropical bursts that were useless to farmers. But they, and De Waal, did not understand back then that this was global climate change. Most people in Darfur felt that God was punishing them.[36]

That misunderstanding is not surprising. It is only in the last two years that the world has begun to understand about climate change in Darfur. But we do understand now. Poor farmers and herders in Sudan have been paying the price for economic development in North America, Europe and Japan. From this follow the only just solutions to the ecological crisis in Darfur.

First, the world should have fed Darfur when famine hit. That food aid should not have gone to refugee camps already full of disease, but been transported by the might of US and Western planes and helicopters to every village in Darfur. But such action on its own will only ever be a temporary solution.

One long term solution can be seen in how Darfuris and people across the Sahel have tried to solve their problem. They have become refugees and tried to get to rich countries. For 40 years the Sudanese economy has only functioned at all because of money sent

Alternative Futures

back by migrant workers in the Arab oil countries. For many poor countries in the world, such "remittances" are much the largest source of foreign exchange. Across the Sahel, climate change refugees have desperately tried to get to Europe as "economic migrants". Some have made it, some have been sent back and some have drowned. What those people need is not cosmetic "aid" programmes. They need open borders. As climate change spreads, the people of the world must insist that we take care of the refugees.

There is another solution too. The heat and sun of the Sahara, Darfur and Sudan make it a prime location for concentrated solar power (CSP). We have already seen that CSP plants in the Sahara could supply the electricity needs of Europe through long distance power cables. That would mean jobs in the Sahara and the Sahel. What Darfur needs is grain, visas and industry, not foreign armies.

All that may now seem visionary and utopian. But the alternative that faces us is for the world to become like Darfur, and Darfur to become something unimaginable.

What has happened in Darfur is history, and that cannot be changed. But there was a moment when the people of Sudan tried to change their history. In the spring of 1985, at the height of the famine, the IMF insisted on still further cuts in the living standards of ordinary Sudanese. An uprising began in the neighbourhoods of Khartoum where the famine refugees were concentrated. They stormed the grain silos. Within days a general strike and uprising had spread across the capital city. The neoliberal dictator, General Nimeiry, fled and the government fell. A feeling of optimism spread across Darfur.[3] People expected that food would come, and a more decent society.

But this was 1985. The right wing and neoliberalism were in ascendance throughout the world. The military dictatorship was replaced by an elected government led by Saddiq Al-Mahdi, an Oxford educated son of the Sudanese elite, from one of the two richest families in the country. Business continued as usual and almost nothing changed.

Other terrible crises will throw up many more such revolts. The hope must be that in future there will be brave and organised networks, parties and movements prepared to lead desperate people. In those moments, they can appeal to the world for help, institute social justice in their own countries, and fight climate change.

CHAPTER 18

Another World is Possible

A WORLD of refugees, famine, war and suffering awaits us if we do not act. Many people say, *The problem is urgent, the planet is in danger, so we have to act now. And given what the world is, we have to act within the limits of the system, within what we can persuade the people at the top to do.*

It's true that the problem is urgent and serious, and indeed we have to be realistic. But there is realism and realism. The truth is that the people at the top cannot and will not solve the problem.

On one level, for most ordinary people, "Don't put your trust in the people at the top" is obvious. But over the last 30 years the people at the top have so dominated the media and public debate that the idea of forcing them or replacing them sounds like breaking a terrible taboo. It is a dramatic and heart-stopping assertion, almost too scary to contemplate. This fear leaves people feeling hopeless and helpless about global warming.

This is the explanation of what people often call "apathy". It is often said that people don't understand the reality of what global warming will mean. So what activists have to do is find a way of scaring the living Bejasus out of people, and then they will act. There is a portion of truth in this strategy. People won't act unless they understand what is at stake. It is necessary to inform the world. But there is also a basic mistake in this strategy. Political apathy is not fun—it's more like personal depression. People do not become seriously depressed because they don't care about their lives. They become depressed because they care all too much, but can't see a possibility of changing their world. That's why depression *hurts*. Political apathy is the state people find themselves in when they know something is badly wrong in the world, but feel helpless to change it.

It's not that ordinary people don't know about global warming now. It's that they don't want to hear too much or think too much about something terrifying they feel they can't stop. On one level or another, most people are all too aware of the scale of changes necessary to stop climate catastrophe. They just don't think that people like them can make those changes.

This produces a sort of split consciousness in many people now. On the surface, they say and try to believe that the politicians can solve the problem. They tell themselves that markets may work. At the same time, in another part of their heads and hearts, they don't think the people at the top will solve the problem. But this very pessimism about politics and politicians feeds their reliance on existing politicians. If I am convinced that we cannot solve the problem, then I have to try to believe that maybe they will.

The solution to climate apathy is not more fear. It is to persuade people that ordinary people can have an effect in the world. For those who believe deep change is necessary, that means political alliances with all the people who think the world cannot be fundamentally changed but who do seriously want to do something.

This is the logic behind every demonstration against global warming. Those demonstrations invite people who feel angry but isolated to march besides others who feel the same, and to learn they are not alone. This is why it is a mistake to say demonstrators are preaching to the choir. The choir in a church sing together, because in those rising voices they can hear their shared faith. Demonstrations make people feel less alone, more powerful and confident. That's why the size of demonstrations matters so much. The more people you march alongside, the stronger you feel. It is also why global demonstrations are so important. Even if you are only 500 people in one town in Senegal in 2007, you know you are part of a global movement, and take heart.

Moreover, a global campaign is necessary because halting climate change will require united global action by almost all the governments of the world. It doesn't have to be all the governments. If one important player is left out—the US, China, India or Japan—they can be disciplined. The other countries could simply refuse to trade with them, and that would bring them into line. But we do need almost all the major powers.

However, it will not work simply to have agreed targets for all the countries. That leaves humanity with only negative targets. It means cutting what we have, not building new and different ways of using energy. Global targets on their own will simply mean generalised sacrifice and therefore won't work.

That's the logic behind all the campaigns I suggested earlier for things like five million solar roofs, ten million insulated homes, the defence of rainforests, car-free cities, building railways and the rest. However, fighting for these things on their own will not be enough. We will also need firm global and national caps in emissions. Without that, banning cars will cut emissions but they will rise somewhere else.

The point in local campaigns, though, is that like demonstrations they can begin to break the problem of "apathy". The key here is to find campaigns that are big enough to make a difference, daring enough to fill people with passion, but still realistic enough to win. That doesn't mean campaigns have to win immediately. Indeed, precisely because they are daring and ambitious, they will take time. But we have to choose campaigns people can imagine winning.

There is a weight of history and defeat in the minds of the living. The new social movements have begun, but only begun, to turn that round. That's why fighting for climate demands that can be won is so crucial. Winning will change what people can imagine.

This is also why all the other struggles of the social movements matter so much. Exasperated climate campaigners sometimes say that they don't see why everyone doesn't just drop all the other causes and fight for the climate now. After all, climate change is the most important threat the world faces today. There is no point, they say, in trying to fight global poverty if climate change is going to plunge billions into poverty.

I know why people feel that way, but they're mistaken. The fundamental problem for climate campaigners isn't that people don't care enough about the planet. It's that they don't think they can fight and win. Every campaign for social justice that fights and wins changes people's minds. And if it's big enough, it changes the minds of tens of thousands or even millions who watch and see that resistance can win.

I have argued earlier that the simple existence of the new anticapitalist movement made it possible for people to think about a global climate movement. Even more important is the slow defeat in Iraq and Afghanistan of the United States armed forces, the greatest power on earth. When withdrawal comes, the world will see the humiliation of the greatest power of all, a power that almost no one thought could be beaten.

As this has begun to happen, we have seen the weakening not just of George Bush and Exxon Mobil, but of all the climate deniers in the world. So the governments of other countries stand up to Washington over climate in ways they would not have done before. And as the American people increasingly reject the neocons and corporations who took them into war, they too have begun to embrace real action over climate.

What we are seeing is only the beginning, however. An actual American defeat, the spectacle of troops scrambling for the helicopters, would change what people believe is possible in the world. The defeat of any government in war always opens up a space for resistance and opposition in that country. The US government has dominated not just the American people but the world.

A US defeat will increase the confidence of climate campaigners everywhere. But that defeat will also encourage people all over the world to mount resistance over thousands of small local issues. Teachers in a secondary school in Liverpool will say maybe we can do something about the way the headmaster bullies us. And villagers in Bengal will feel braver and find it easier to stop the state government seizing their land and handing it to corporations. Those thousands of small and tiny struggles will feed back into a global shift of feeling that, yes, we can change the world.

However, all this will only happen *if* and *when* the US withdraws from Iraq and Afghanistan. That's why the global peace movement, and especially the American peace movement, matters to the future of the climate. People who can stop war can stop climate change. People who can't make their governments make peace won't have the courage and confidence to make them save the climate either.

There's another reason why all the social justice struggles against neoliberal policies, in particular, are important to climate politics. The serious solutions we need to cut emissions all fly in the face of neoliberalism. They mean government spending, public works and government regulation. People will imagine that is really possible only if they can defend the public services they still have, and begin to win new ones. When the new government of Bolivia nationalised the foreign run gas fields, I watched it on British television. President Evo Morales issued a decree, and that same night he marched into the gas fields with the armed soldiers behind him, all shining in the harsh spotlights. I *saw* that it was possible to for human beings *now* to take control of their own economy. So did millions of Bolivians, and people far beyond Bolivia. More victories like that, many more in many places, will give people courage to do what is necessary to save the planet. But only if those struggles happen, and only if they win. That's why the myriad fights against neoliberalism matter for the climate and why it would be madness to campaign only about climate change.

However, what I'm definitely not saying here is that social justice activists should take care of their own struggles while environmentalists defend the climate. We simply won't be able to stop climate change without social activists, trade unionists and the left, and much wider forces than them. The sheer scale of the movement we have to build means that environmentalists can't do it on their own. A mass climate movement has to include and mobilise large numbers of working class people. Not all workers are in unions by any means, but the easiest way to mobilise workers is still through the unions.

Moreover, the ideas of social justice will be critical to the success of any climate movement. That means the left has to be in the climate campaigns. It can be all too tempting for the left to stand on the sidelines and explain that capitalism causes climate chaos, and only revolution can save the planet. Those things may be true. But it is useless for the left to make formally correct arguments in isolation.

The converse is also true. An environmental movement that does not fight for social justice will be unable to defend the environment. The left and the unions have to join the climate fight. But the environmentalists must also go to them and insist that they do so.

This has already begun to happen, in both directions, in one country after another. This book, indeed, is only possible because that is happening.

Can we do it?

None of this means it will be easy to stop climate change. But people have changed the world before. The world was once run by kings and feudal lords. It is now run by business people. The "bourgeoisie" did not change the world by inches; they led whole populations in revolt. It began on the edge of the world system, with the American Revolution of 1776. The revolution of 1789 in France, at the centre of the world system, was decisive. And then it spread across much of Europe, and all of Latin America.

In 1919 the movement for colonial independence exploded with Congress in India and the May 4th demonstrations the same year in China. By 1948 India was independent, and by 1949 the Chinese revolutionaries had won. Now colonial independence has been won almost everywhere in the world, with a few glaring exceptions like Palestine and Tibet.

There was a time when ordinary people did not have the vote. In the 20th century great movements fought for that. Now people vote in most places, and the people who do not long for democracy and know the dictators can be overthrown. People fought to build unions, and now the big industries and the public sector in most countries have unions. People fought for the welfare state—free schools, hospitals, money for the unemployed and disabled, and pensions. Now most people in the richer countries have those things. Of course the institutions of the welfare state are flawed, and governments are now trying to take them away. But healthcare, welfare, education and pensions are still things people fought for and won, and they have made life infinitely better for billions of people around the globe.

I know great movements for change are possible. I grew up in Texas in the 1950s and 1960s, and I was part of the civil rights movement that changed my world. I was also part of the movement that stopped the Vietnam War. Women's liberation and gay liberation have not won perfect equality for women and gays, but they have changed the lives of every woman, every gay man and every lesbian in the world.

The scale of the task that faces us now, to save the planet, is of the same order as these great social movements. The scope of the battle lies somewhere between the French Revolution and the fight for the welfare state.

We know we can change the world, because people like us have done it before. We are not genetically different from our fathers and mothers, our grandparents and our ancestors. It can be done.

Capitalism

But which is it? Is climate change like the welfare state, or is it like the French Revolution?

To answer that question, we have to start with the underlying problem—capitalism. After all, the system of industrial capitalism has grown up hand in hand with the carbon economy. It started with coal power and the steam engine. Then industry moved on to electricity from coal, transport based on oil and heating from gas. As coal, oil and gas were dug out of the ground, they allowed an unprecedented expansion in the use of energy. The last 200 years have seen a doubling of life expectancy, a six-fold increase in the world's population, and an even greater increase in the total production of the world. Capitalism now dominates the world, and every piece of that world system now depends on carbon fuels.

So in one simple sense capitalism is the cause of global warming. This does not mean that capitalists and corporate executives set out to cause climate change. They had no idea such a thing would happen. It was an unintended, and unwelcome, by-product of the dominant technology. But it does mean that the capitalist system now stands in the way of the changes we need.

In earlier chapters I have outlined why this is the case. First, the entrenched policies of neoliberalism rule out the massive public works and government regulation that are needed. Second, many of the most powerful corporations on the planet are inescapably tied to carbon.

Neither neoliberalism nor carbon corporations are essential to capitalism as a system. After all, 40 years ago capitalist corporations coexisted happily with nationalised industries and government intervention and regulation. Back then corporate and political leaders favoured Keynesian policies. There is no iron law that says they could not do so again. But there is a stubborn economic reality. The corporate world swung away from Keynesian politics because of a dramatic fall in industrial profits in the 1960s. The rate of profit in industry has since recovered only partially. Corporate leaders and mainstream politicians may well feel that they cannot now give up squeezing working people, whatever the threat of climate disaster.

It is also possible to imagine a capitalist system that makes healthy profits from public transport, wind turbines and solar power. But the carbon corporations are very powerful, and time is short. The car and oil industry may stall long enough to destroy the planet.

In short, neoliberalism and carbon corporations are not essential parts of any capitalist system. But they may well be so powerful, so entrenched, that we will not be able to act in time without changing the whole system.

Moreover, the capitalist system throws up two further roadblocks—global competition and relentless growth. Until now international negotiations for a new climate treaty have been characterised by the intransigence of Washington. But this is only one possible face of global competition. If American power wanes, or the American people prevail upon their government, we will still be left with the reality of competition within capitalism. That will mean other countries, in other combinations, will stand in the way of progress.

More important will be the relentless pressure for growth in each part of the capitalist system. The corporations and nations that profit, invest and grow are the ones that survive. The pieces of the system that do not grow do not simply stagnate. They lose the race, fall by the wayside and wither. This is the logic of the system. It is part of the reason why governments like the British seem so

　　　　　　　　　　　　　　　Alternative Futures

hypocritical now. New Labour call for global cuts in emissions. They will pass a climate bill that demands annual cuts in emissions. They say they are making such cuts right now. Yet emissions in Britain, and most other places, grow relentlessly. And the New Labour government's policies also promote massive road building, new airports and new runways, new coal fired power stations, and plans for endlessly growing energy use. This is lying and hypocrisy. But it is not simply that. It is the living contradiction between the human necessity of halting climate change and the capitalist necessity of growth.

There is a further problem with the dynamic of growth. Without growth the system will wither. Yet the current global growth rate of only 3.5 percent will mean a doubling of global production in 20 years time, a quadrupling in 40 years. In the long run that means an inexorable rise in manufactured things, and therefore in greenhouse gas emissions.

The good, and important, news is that growth does not need to mean a rise in emissions in the next generation. The capitalist system needs an expansion in investment and production. Gross national product has to grow. But gross national product is only a measure of economic activity—of total wages and prices. If the world is covered with wind turbines, solar panels, concentrated solar power plants, and tidal barrages, all that will count as economic growth. So will the insulation of buildings and the building of new electric railway lines. For a generation we can have economic growth of a new kind and drastic falls in carbon emissions. For the moment we can worry about that one later.

However, we need to face the facts. The rulers and defenders of global capitalism will be under enormous pressure to stand in the way of the measures necessary to stop climate catastrophe. This will be true even when they come under immense pressure to act for the planet.

It is possible to imagine four possible outcomes of those conflicting pressures. The first outcome is that the defenders of the present order stall for long enough to ensure abrupt climate change overwhelms us. This is all too possible.

The second outcome is that the rulers of the world come to their senses and take at least partial control of their system. This would be welcome, but is the least likely outcome.

The third outcome would be a mass movement so strong and determined that it can force the rulers of the world to act. That would mean large compromises between current capitalist reality and human need—deep and lasting social and economic change. In a sense, the powers of capitalism would be forced into action against their better judgment. This is more likely than a simple change of heart from the top. Something rather like that happened with the fight for welfare state in Europe in the 20th century. Socialists, many of whom thought the welfare state was incompatible with capitalism, fought for it anyway. For all the limitations, they did win pensions, unemployment benefits, aid for the disabled, social welfare, public housing, government run public transport and national health services. All this happened under capitalism.

A compromise of that nature will not come easily. The switch to a renewable economy requires even larger changes in the system than the coming of the welfare state did. I don't think it's going to happen. But it would be foolish to rule out the possibility. In any case, that kind of compromise, that kind of deep and lasting change, will not happen without a massive global movement that is prepared, if necessary, to change everything to save the planet.

The fourth possible outcome is that the ordinary people of the world take control of their societies and economies. This will mean revolution, a revolution in the hearts and minds of human beings, but also an overturning of global corporate power. It would be another world, a true global democracy. That may not be the most likely outcome. But it is possible, and it would save the planet.

What about human nature?

As soon as I say in a meeting we have to change the world, someone usually sticks up their hand and says that human nature means that people are too greedy and individualistic, and so we can't save the world.

Is that true? The answer is that there's a lot of evidence both ways. We have all seen enough of life to know that greed, cruelty and self-obsession are all around us. They are a part of human nature.

But there are also other parts. Greed and individualism seem normal to us because we live in a competitive capitalist society. Ours is a world where everything seems to be for sale. Individualism is the personal form of that. Capitalism is driven to constant accumulation of profits. Greed is the individual expression of that. Competition rules in capitalism, and we value achievement and winners. Our dominant virtues are the virtues of businessmen.

In other times and places other characteristics have seemed more normal and ideal. In the Middle Ages in Europe power lay with

Alternative Futures

knights and lords—thugs on horseback. The dominant and natural human virtues were courage, loyalty and physical strength—warrior values.

In the 1960s the anthropologist Richard Lee lived with !Kung San hunters and gatherers in the Kalahari Desert in Botswana. Their values were different again. It may have been because they were hunters, constantly on the move, and owned no more than they could carry on their backs. Or it may have been because they were also modern people right at the bottom of the social hierarchy in Botswana. In either case, their values emphasised sharing because they had to share to survive. They valued equality, and a man who was not pushy. And when relations got too tense, in families or in camps, people simply moved away for a while.[38]

We see the same mix of values in any modern industrial society. The corporate world at the top is a circus of competition and greed. But most people try to raise their children with love and kindness. Indeed, without caring and sharing much of the work of teaching, nursing and cleaning would never get done. Without the quiet kindness of colleagues, work becomes a hell. This does not mean that either solidarity or competition is more human. Not all parents treat their children with love and respect all the time. Indeed, every parent has felt those moments when your hand rises and love and rage are at war inside you.

In short there is no one human nature. There is evidence for many possible humanities.

Another world

A lot of people I meet now have figured out for themselves that we have to change the world to stop climate change, and that means a revolution. But there, often, they stop, because they can't imagine what a revolution, or another world, would look like.

I can't know for sure either. The future lies in front of us, and is unknown, and will not be a simple repeat of the past. But I have some guesses.

Let's begin with the idea of "revolution". We have that word because people have made revolutions before. Each was different— the American Revolution of 1776, the French Revolutions of 1789 and 1848, the Russian Revolution of 1917, the Chinese Revolutions of 1910 and 1949, the Iranian Revolutions of 1906 and 1979, the Mexican Revolution of 1910, the Nepali Revolution of 1990, and many more. All of them, though, have some things in common. Sometimes the revolution led to war. But always there was a mobilisation of the people. There were strikes and fervent meetings and parades through the streets into the teeth of official gunfire. There were pamphlets, speeches, passion, endless talk and hope. Always people started with small ambitions—they asked the king for only a little change, a little justice. And they ended up toppling the king and changing their world.

When the revolutionaries won they did so because, when it came to a choice, the great majority chose the revolution over the old order. Indeed, the revolution was that majority in action. But the initial uprising is only half the story. The other half is what comes afterwards. Indeed, in the last 20 years insurrections have become common across the world. The tragedy is the regimes that have replaced the old order in Nepal, Sudan, East Germany, Czechoslovakia, Nicaragua, Thailand, Somalia, Afghanistan, the Phillipines, Iran and many more countries.

The real question is what we mean by changing the world. To answer that, I'll start with what we can do to stop climate change.

First, any revolution must be global because we are all locked into one global economy. Any revolution left isolatied withers into defeat or military dictatorship. But that isolation does not have to happen. All change begins in particular places. But the world is very similar in each of its parts these days. The political ideas, the debates, the fashions and the structure of feeling have more unity than ever before in history. If one important country changed utterly, everyone would notice, and the example would spread like a forest fire.

This is even more important for climate change. We share one atmosphere. Change must start in one country, but it is pointless if it does not spread to all.

Second, ordinary people have to take control of the economy. The dynamic of profit has to be broken. This is easier than it sounds. Think of primary schools. Private schools are run to make a profit. Public state schools are run to teach children. That's what they are for. Even now governments decide how much to spend on which schools and what they want them to do.

Of course, neoliberal governments are now furiously trying to privatise schools, to make them compete with each other and to force them to make paper "profits". The very fact that they are forcing this through, though, is proof that schools have long been run in another way.

Alternative Futures

With a revolution we could treat the whole economy that way. The bottom line would not be profit, but need. So we would ask, collectively, do we need more solar power? More old people's homes? More steel? More holidays and less furniture? Less packaging? More meat or less? More flowers or more water? Ads or live music?

We would also decide collectively how to share out the wealth of the world. Do we give doctors more than slaughterhouse workers, because they have more education? Or the same, because everyone deserves an equal share? Or less, because doctors get more satisfaction at work?

Then there are the inequalities between regions and countries. Do we try to equalise wages and incomes across the world immediately? Or do we hold Japanese and American wages steady, and wait for Bengali and Sudanese incomes to rise?

If we break the power of profit, then we can produce for need. That means for the needs of human beings. But it also means we can take care of the atmosphere and the environment. We can care for children and old people, but also for oaks and salamanders. Humanity now bestrides nature, uses, exploits and "conquers" nature. We can be returned to the feeling that we are part of nature.

If we break the power of profit, we can also break the relentless need for "things". In an early chapter I talked about why people need things—commodities—now. It is not that they make a person relaxed and healthy: this is much more a result of their standing in society. The things are only a measure and a sign of that standing, but because they are that measure, people cling to them.

But imagine now a world with a global economy that is not driven to invent advertised needs. One where people do not need to count and display their things to measure themselves. Where indeed measurement may not be the main concern or business of humanity. Where everyone has enough. Where social benefits insure that no one will be homeless, jobless, bereft and desperate. Where the choices are not between bling and destitution.

The global economy now is saturated with waste. Not just plastic bags and throw away tins, but whole industries. Advertising, where millions of artistically gifted people throw their lives away. Security and ticket taking, where people are bored out of their minds policing inequality. The layers of managers in public services, filling in forms and ticking boxes.

With the end of the tyranny of profit and growth, people could ask together, do we need this? And we will find that what we need is not what we need now.

None of this means sacrifice. We live in a world of fear now. For some it is the fear of soldiers and the threat of violence. But for most it is the fear of loneliness in old age, the fear of being humiliated in school, by the boss at work, by the bored and angry person who guards the door. We live in a world where people are told what to do and used by other people to make profits.

To live without that, to live with enough—that's what we can have if we end the tyranny of profits. And without that tyranny we can fit ourselves to the land.

Power and democracy

Well, yes, some might say. That sounds nice. But you know it won't be like that. If everything is public, there will be no freedom. Everything will be repression, like the old Communist states.

But have a look around you. Watch Baghdad burn. Remember the man in the hood with the electrodes running from his hand? He is legion. See if you can hear the cries from Congo, where three million people have died in a war fought for control of a valuable mineral that is absolutely essential to the manufacture of Play Stations. We live in a world of repression.

None of this justifies those old Communist states, with their tanks and secret police and prison camps. Those obscene regimes began with the noblest of human dreams. Back in 1917 the friends and enemies of the Russian Revolution all round the world knew the workers had taken power there. Then most of the powers in the world invaded Russia, blockaded the economy, and armed and supplied an army of counter-revolutionaries. When that was done and something of the revolution still survived, the country was broken by poverty. Around it hostile powers armed. The Russian state, isolated, used and squeezed its workers and peasants to build a war machine that would be able to withstand German invasion. That they did, and lost all public decency in the process, and the government had to kill almost every single activist in the revolution, for they could not be trusted to live and talk in this new brutality.[39]

Could it be like that again? Yes, even with the highest hopes, if a revolution is poor and isolated. But it doesn't have to be like that, if we change the world, not just one country.

The third thing to say is this. If we don't change the world we are going to face abrupt climate change. In that shifting kaleidoscope of disaster, lives, homes and societies will be devastated. The rich and

Alternative Futures

the corporations will know that someone has to sacrifice, and in no small measure, and they will kill who they need to in order to make sure it is not them. That inequality, that lack of public caring, will demand a routine repression that people only face these days in the worst places. And they will go to war.

But the last thing to say is this—the only safeguard is democracy. The vote, yes. But not just votes every five years, choosing between lawyers who are going to betray you. Votes everywhere, at work for the boss, in the neighbourhood, in school, in the old people's home, in the kindergarten and on the train. Votes to make the global decision about what is made, and how, and what people will have.

There is only one guarantee of that democracy—us. We will not change the world unless in the end the great majority of people in the world come together to do so. It is hard to imagine that you and I, and the people we know, could do that. For a revolution is not only a change in the mechanisms of society. It is a change in the human soul. And it is a change in every relationship in society. It is the moment when people refuse to be bullied, and agree to stand by each other.

Climate justice and social justice

Is all that possible? Yes. We could do that. Is it likely? I don't know. That's down to you.

But maybe we don't change everything, beautiful as that will be. Maybe all we need is a great social movement with a vision, and we can force the powerful to compromise. Maybe we can save the planet anyway.

I hope so.

For now, and for the foreseeable future, any serious climate movement will have a large majority who think revolution is not possible. If you look around your movement and see a majority of revolutionaries, that's just evidence you are isolated from society as a whole.

But a much larger number of people will want to dream of what they think is not possible. Never underestimate the power of that dream, the courage people can draw from that vision, to keep slogging on in the world we're in.

I have two last points to make. The first is about the relationship between stopping climate change and power in society. People often talk as if there was a simple choice between working within the established limits of society and opposing them. In reality, it's not like that. Revolutions don't start because people decide to change everything. They start often because large numbers of ordinary people want to preserve what they have. And as they try stubbornly to do that, they find they have to change the world.

When people try to change the world, even in seemingly small ways, they find themselves launched into struggle. Over the years I've walked to a lot of picket lines where workers were on strike over pay. I asked questions and listened to the answers. Over and over, the issue is pay, but what people are talking about is the look on the foreman's or the supervisor's face. When they consider defeat, they tell me how hard it will be in there, that the manager will be able to do anything to them. When they win, they tell you they're going in with their heads held high, and the managers are looking anxious.

Every struggle in society shifts power, even if slightly. The debates over climate policy will be not be straightforward and consensual. If we fight for and win deep cuts in emissions, or car-free cities, or ten million solar roofs, those will not simply be policy changes. They will also be major shifts in power and confidence within the society as a whole. The corporations and the ruling politicians will be weaker, the activists and the people stronger.

That's why it makes no sense to think of simply changing climate policy and leaving the rest of established society the same. If people have the power to change what they do to the air, they will have more power to change the rest of their lives too. The rich and powerful know this. If we even manage to challenge the priorities of capitalism, there will come a time when the rich and powerful will feel people power must be stopped, even if the world has to burn.

What is at stake

Finally, there is much at stake. We face, in this generation, a turning point in history.

For hundreds of thousands of years people lived by hunting, fishing, gathering wild plants and burning wood. Our nature as social animals allowed us to live reasonably well with others of our kind. Little bands of foragers had to be constantly on the move. No person owned more property than they could easily carry easily on their back. Anyone who was unhappy with their lover, neighbours or parents could simply move to live with relatives in other bands. There

Alternative Futures

was still some murder, and probably some warfare for hunting territories. But there was no unequal access to food and resources within the band. We took care of each other, more or less.[40]

Then about 12,000 years ago people invented farming. That gave humanity an enormous increase in food and population. What we had done, with our hands and our brains, made a million new possibilities. But people were now tied to the land, waiting for the crops. Property was fixed. Inequality grew, and with that slavery, lords, kings, violence and war. Ordinary people dreamed of equality and spoke of gods who loved the poor. Peasants and slaves rose up against their masters over and over again.

The next great change began with the industrial revolution two hundred years ago. That change was based on science and factories. The new system burned oil, gas and coal. Again there was a great leap in food and population. People lived longer and fewer babies died. Education flourished and people learned to read. There was no need any more for anyone to live in want or fear. Again people struggled to build a social system that could share the wealth. Instead inequality, war and cruelty grew to previously unimaginable heights. Capitalism, the present economic and social system, has given us antibiotics, enough food for the world, and a global culture. But it has also given us Auschwitz, Hiroshima and the Ethiopian famines. In the new system, capitalism, people fought against the masters again, though they talked less of gods. They won the vote, and in some rich countries the welfare state. When they thought they had fought and won everything, in Russia or in China, it turned out they only had new masters. But still they dreamed.

We are animals, but animals of a new type. Without our choosing, our hands and brains have bestowed upon us the stewardship of all life on Earth. We face a moment of choice. Since we discovered farming and industry, the work of our hands and brains has raced ahead of our ability to create a society fitted to the new technology. We are now animals with nuclear weapons. The question is what we will become. Global warming makes that choice acute and urgent.

NOTES

Introduction

1 The numbers come from the list of the 500 largest global companies by *Fortune* magazine. See www.money. com/magazines/fortune/global500. These are the ten largest corporations measured by sales. This measurement gives the best estimate of political and economic weight, and of share in gross national product.

2 See Timmons Roberts and Parks, 2007, for the scale and numbers of dead from climate disasters in the Global South.

Part One

1 Estimates of how many species will disappear in abrupt climate change vary widely—no one really knows.

2 The two best introductions to the science of climate change are Pearce, 2006, and Flannery, 2005. For good explanations of abrupt climate change go to Pearce, 2006; Flannery, 2005, pp189-205; and Alley, 2000. Cox, 2005, and Mayewski and White, 2002, are also very useful.

3 I take this very important point from Pearce, 2006, pp300-301.

4 We know emissions are falling only because the amount of methane in the air is falling slightly. Otherwise, it is difficult to measure methane emissions. Carbon dioxide emissions come mainly from burning coal, oil and gas. Every government keeps statistics for how much of those fuels are used. But methane emissions come from diverse sources that are harder to count, and estimates vary greatly. The higher estimates of human emissions are five times the lower estimates.

5 Pearce, 2006, p192; Cox, 2005, pp113-120; and Alley, 2000, pp111-112.

6 Cox, 2005, pp129-144.

7 Flannery, 2005, pp196-199.

8 Start with Hansen and others, 2007.

9 Pearce, 2006, pp62-81.

10 Pearce, 2006, pp109-112.

11 Benton, 2003, p272.

12 Stern, 2007.

13 See Bows and others, 2006, pp163-166; Stern, 2007, p228; Intergovernmental Panel on Climate Change, *Climate Change 2007: The Physical Science Basis*, 2007 (see "Technical Summary, section TS.5").

14 When scientists talk about all the greenhouse gases together, they use the idea of "CO_2 equivalents". For instance, there are now 385 parts per million of CO_2 in the atmosphere, but there is a total of about 435 parts per million of CO_2 equivalents. For most of the rest of this book, I will simplify things by focusing only on CO_2 levels.

15 The figures for emissions come from the US Energy Information Administration, both from their *International Energy Annual* and from the spreadsheets available at www.eia.doe.gov/pub/international/ca rbondoixide.html. The European number is a bit misleading. It includes all European countries except the former Soviet Union. That means it includes some small countries in Eastern Europe that bring the numbers down. The average for Western Europe is 8.7 tons per person.

16 Wen and Li, 2006, pp140-2.

17 The figures on fertility are from Population Reference Bureau, 2006 World Population Data Sheet, available at www.prb.org.

18 The Population Reference Bureau gives the official figure for China as 1.6. This is low because of the government's repressive one-child policy, and hundreds of millions of forced abortions, often in the last month or two of pregnancy. However, resistance is widespread and many Chinese children, particularly girls, have been hidden from the official statistics. To allow for this, the Chinese government estimates that the real number is 1.8. Reliable demographers think it may be more. Also the one-child policy is slowly breaking down in the face of popular resistance, especially among rich people who are now often willing and able to pay any fines. So the real number is probably somewhere around 2.0 children per woman. This is a guess. It may be a bit less or a bit more. For the one-child policy on the ground, see particularly White, 2006, and Greenhalgh and Winkler, 2005.

19 Krause, 2006, is interesting on this.

20 For population see Rao, 2004: Foster, 2000; and Mamdani, 1975.

21 The high and medium estimates depend on unrealistic assumptions. The medium estimate assumes that population growth in the poor countries will fall in ways that replicate the patterns other countries have gone through. This seems reasonable. However, they also assume that the rate of children per woman in all countries will not go below 1.85. They also assume that where the rate is less, as it is in most rich countries, the rate will come back up to 1.85. There is no good reason to make these assumptions, and they make the medium estimate unduly high, let alone the high estimate. A second reason for rejecting the high estimate is that for the last 25 years all official predictions of population growth have had to be revised downwards, because they were too high.

22 These population predictions, and the ones that follow, are from the Population Division of the Department of Economic and Social Affairs of the United Nations Secretariat, *World Population Prospects: The 2006 Revision*, accessed at http://esa.un.org/unpp.

23 Emissions were 27 billion tons in 2004. Double that to 54 billion tons in a world like Italy. Add 15 percent for population growth. That makes 62 billion tons. To reduce that to 11 billion tons is an 83 percent cut. To reduce it to 8 billion tons is an 87 percent cut.

24 For the lives of agricultural labourers in India see Breman, 1996 and 2003.

25 For Chinese workers in the Cultural Revolution and now, start with Perry, 1995, and Ngai, 2005.

26 See the 2005 *Annual Report* of the US Congressional-Executive Commission on China, at www.cecc.gov/pages/annual Rpt/annualRpt05/. The statistics are published each year (in Chinese) in the official Chinese government magazine Outlook. For a fascinating analysis of the protests and strikes, see Gilbert, 2005, available at www.isj.org.uk.

27 See www.pewglobal.org/reports for 27 June 2007.

28 McGregor, 2007.

29 Navarro, 2006.

Part Two

1 Koistinen, 2004, p276-277. This section relies on Koistinen.

2 Lankton, 1991, p42; and Koistinen, 2004, p254.

3 Koistinen, 2004, p288. As Koistinen says, this is an unadjusted figure. It does not allow for the fact that much of this new value could not be used after the war. He suggests that the long term increase in value of plant was between a quarter and a third.

4 Koistinen, 2004, p430. I can't tell from Koistinen's text whether the rise in corporate tax was a rise in the basic rate, or in the percentage of profits actually going in tax.

5 Koistinen, 2004, p477 for profits and p438 for incomes. Immediately after the war family incomes fell back by 4 percent after tax, but that was still a 43 percent increase overall in seven years.

6 Koistinen, 2004, pp344-345.

7 Koistinen, 2004, p251. The total miles travelled also increased, so this is a more than fourfold rise. Rail freight also doubled.

8 Personal communication, Ernie Roberts.

9 Zweiniger-Bargielowski, 2000, pp137-140 and Burnett 1979, p330.

10 For electric cars see Paine, 2006.

11 The discussion of clean energy that follows owes much to Monbiot, 2006, pp79-141.

12 The standard, comprehensive and wonderful source on wind is Gipe, 2004.

13 The cube of the wind speed means that you multiply the wind speed by itself twice to get the amount of electricity produced. So let's say that the wind speed is 10 miles an hour. That will produce 10 times 10 times 10 units of electricity. That's 1,000 units of power from 10 miles per hour (mph) of wind. Double the wind speed to 20 mph. Then multiply 20 times 20 times 20. That gives you 8,000 units of electricity—eight times the power for twice the wind speed. The power is also affected by the size of the circle of wind the blades sweep through. In

Notes

technical terms, the power increases with the square of the length of the propeller blade. The square means you multiply the number by itself once. So with a blade 10 metres long, you get 10 times 10, or 100 units of electricity. Make that blade 50 metres long. Then 50 times 50 is 2,500 units—25 times as much power. A long propeller blade means a high wind turbine. So a whacking great turbine in a windy place is enormously efficient. Now think about the effect of doubling the wind speed and making the turbine blades five times as long. That gives you 8 times 25, which is 200 times as much electricity.

14 Monbiot, 2006, p131.

15 For a fascinating read on one such battle see Williams and Whitcomb, 2007.

16 For some wisdom about good design on wind farms, see Pasqualetti, Gipe and Richter, 2002. Particularly chapters by Gipe, Pasqualetti, and Nielsen.

17 Property owners who don't want wind farms also say the turbine blades kill birds. This seems to them another environmental argument. And it's true, wind turbines in some places do kill birds and bats—usually migratory birds and bats, not local ones. But they kill fewer birds than electricity. More important, global warming won't just kill some birds; it will wipe out whole species. For more on birds and bats see Gipe, 2004, pp298-301, and Sustainable Development Commission, 2005, pp65-71 and 153-166.

18 Archer and Jacobson, 2005.

19 For solar power, see Bradford, 2006; Scheer, 2002; Scheer, 2005; Leggett, 2005; and Monbiot, 2006, pp100-142.

20 Bradford, 2006, p9.

21 Scheer, 2002, p99.

22 The following account of Japanese and German PV is based on Bradford, 2006, pp99-109 and 178-181.

23 Rabe, 2003, and State Energy Conservation Office, "Texas Wind Energy", at www.seco.cpa.state.tx.us/re_wind.htm

24 The main material used in PVs at the moment is silicon, though others are possible. The companies building PV cells in Japan soon discovered that their engineering problems were very similar to those faced by computer companies trying to get the maximum out of their own silicon chips. Computer and electronics companies rapidly came to dominate the solar power industry in Japan.

25 One state—Washington—used the German-style feed-in tariff.

26 Bradford, 2006, pp180-182.

27 Monbiot, 2006, pp106-106.

28 Monbiot, 2006, p106.

29 Monbiot, 2006, p104.

30 German Aerospace Center (DLR), 2005 and 2006.

31 German Aerospace Center (DLR), 2005, p157.

32 German Aerospace Center (DLR), 2005, p55.

33 Monbiot, 2006, p108.

34 Barrett, 2006.

35 The phrase "rewire the world" comes from Gelbspan, 2004.

36 The California law came into force on 1 January 2007, and budgets $3.3 billion for a million roofs by 2018.

37 Hickman, 2007.

38 These figures are extracted from figures in Intergovernmental Panel on Climate Change, *Climate Change 2007: Mitigation*, 2007, Chapters 5, 6 and 7. The numbers are not exact, and for our purposes here the important thing is the rough proportions between different end users. The figures for the proportion of end use by heating, and lighting and appliances are the least reliable. I created them by assuming that the proportion in most rich countries was the same as in the US, and in poor countries the same as in China. Neither assumption is true. But it is still undoubtedly the case that heating is the largest single end use, and lighting and appliances are the next largest use in buildings.

39 For the USA see the table on "Total energy-related carbon dioxide emissions by end-use sector" at website of US Energy Information Administration, www. eia.doe.gov; and for the UK, *Decarbonising the UK*, p31.

40 Intergovernmental Panel on Climate Change, *Climate Change 2007: Mitigation*, 2007, Chapters 5, 6 and 7.

41 This figure is rough. In the USA the split in 2005 was 56/44 (US Energy Information Administration, "Total Energy-Related Carbon Dioxide Emissions by End-Use Sector"), but the US is probably particularly commercialised.

42 Monbiot, 2006, p65.

43 The discussion of air conditioning that follows relies on Roaf, Crichton and Nicol, 2005, particularly pp217-268.

44 Email from building physicist Fergus Nicol of London Metropolitan University, the world's leading radical expert on air conditioning.

45 Roaf, Crichton and Nicol, 2005, pp247-249.

46 This would, however, be difficult to do overnight. One recent study in Britain looked at existing lighting systems in people's houses and found that in 60 percent of cases they could not directly replace the bulbs for various reasons (Fergus Nicol, email communication). So people and landlords would need time to change their light fittings. In the meantime, we could make it illegal to manufacture incompatible fittings.

47 Assume that 300 tons of CO_2 emissions for generating electricity do the same job as 100 tons of CO_2 from heating oil. Then assume that two thirds of total electricity generation comes from clean energy. That brings electricity emissions down to 100 tons, equal to the heating oil, and no net gain. But assume that 90 percent of the electricity supply is clean energy. Then the CO_2 from electricity falls from 300 tons to 30 tons. This is 30 percent of the 100 tons from heating oil, a saving of 70 percent.

48 Intergovernmental Panel on Climate Change, *Climate Change 2007: Mitigation*, 2007, Chapter 5.

49 From the World Business Council for Sustainable Development, 2004b, quoted in Intergovernmental Panel on Climate Change, *Climate Change 2007: Mitigation*, 2007. These are numbers for fuel usage, not carbon emissions, but almost all fuel usage in transport is petroleum based. My category of cars here is actually "light-duty vehicles" in my source.

50 Using several different sources, it is possible to get a rough idea of the savings in CO_2 emissions from using public transport.

Daniel Sperling and Deborah Salon did calculations for "carbon dioxide equivalent emissions" [CO_2e] per passenger kilometre in developing countries in 2004 (see Sperling and Salon, 2002, p15). Taking their middle range guesses, the differences are striking:

Estimated CO_2e emissions in grams per passenger mile in developing countries, 2004

Car with one person	375
Car with 2.5 people	150
Minibus with 12 people	55
Train, 75 percent full	35
Bus with 40 people	25

On these estimates, a change from cars with 2.5 passengers to buses and trains would cut those CO_2 emissions by 80 percent.

Another estimate comes from putting together a range of British government statistics. (See Bows and others, 2006, p39). These are figures for CO_2 emissions per passenger kilometre and average passenger load in Britain in 2004. (The figure for emissions is taken from fuel use—it is tons of oil equivalent per thousand passenger kilometres):

Comparison of passenger emissions, UK, 2004

	passengers per vehicle	emissions per passenger
Cars	1.6	37
Buses	9	28
Trains	93	10

On these numbers, a bus passenger saves about 25 percent of emissions and a train passenger about 73 percent. However, the bus occupancy rates for Britain are very low by European comparisons. The average British bus carries 9 passengers. The average Belgian bus carries 32. A British train carries 93, a French train 183. If Britain filled the same number of seats, the savings would be 79 percent on buses in Britain and 86 percent on trains.

Another study estimates that Polish car journeys emit about three times as much as a minibus, four times as much as a train or long distance bus, and eight times as much as a local bus (see Mieszkowicz). On these estimates a switch to trains and buses in Poland could cut emissions by over 80 percent.

As a final demonstration of the advantages of public transport, Canadian government statistics from the 1990s gave the following figures:

CO₂ emissions per passenger mile, Canada, 1990s

Car, actual load of passengers	146
Train, actual load of passengers	92
Bus, actual load of passengers	76
Train, full	47
Bus, full	23

(Accessed at Environment Canada, National Environmental Indicator Series: go to www.ec.gc.ca/soer-ree/English/Indicators/Issues/Transpo/Tablees/pttb04_e.cfm.)
On the Canadian figures, a switch from cars to buses and trains would cut those emissions by about 40 percent. A switch to full buses would mean a cut of 85 percent.

51 See the website of the World Carfree Network, a very useful group, at www. worldcarfree.net/

52 For a development of these ideas, see Monbiot, 2006, pp146-154.

53 Truck bodies can be made from lighter carbon fibre instead of steel. Other major changes now possible include improved tires and brakes, and hybrid engines like those already found in some cars. Further changes are possible in aerodynamics, important for trucks because they ride so high. The solution is partly to reshape the lines of the trucks, so they look more like planes. It also partly to reduce top speeds. With cars and trucks, like planes, the effect of aerodynamic drag increases with the square of the speed. The effect is that drag at 70 miles per hour is twice the drag at 50 miles per hour. This means fuel use per mile is much less at 50 than at 70. A speed limit of 50 would have other good effects. Much of the engine size and strength is required for the

ability to accelerate that last 20 miles an hour. Reduce that, and you have a smaller engine and smaller brakes. That means less emissions again, and a lighter truck.

54 Penner and others, 1999.

55 This is achieved by using plastics instead of aluminium and reducing the weight. Airbus is developing similar technology—Clark, 2007.

56 Bows and others, 2006, p40. As airplanes last 30 years, these savings would involve destroying current planes and replacing them with new ones, an expensive business.

57 This could be linked to the fight that is already happening in hundreds of towns and cities to stop Wal-Mart from opening new stores. See Greenwald, 2006.

58 For more information go to www. climatecamp.org.uk

59 Murray, 2001, and Wolmar, 2005.

60 Intergovernmental Panel on Climate Change, *Climate Change 2007: Mitigation*, 2007, gives a figure of 37 percent.

61 The figures are from the Intergovernmental Panel on Climate Change, *Climate Change 2007: Mitigation*, 2007. The figure for petroleum refining is the least reliable. Estimates are that refining and associated processing of petroleum take 5 to 7 percent of global primary energy. I have rather arbitrarily equated this to 6 percent of global emissions.

62 Intergovernmental Panel on Climate Change, *Climate Change 2007: Mitigation*, 2007, Chapter 7, p461.

63 See, for instance, Lovins and others, 2004; Hawken, Lovins and Lovins, 2000; and Weizsacker, Lovins and Lovins, 2001.

64 Intergovernmental Panel on Climate Change, *Climate Change 2007: Mitigation*, 2007, Chapter 7, gives a figure of 65 percent for the EU-25 and 63 percent for the US.

65 Tilman and Hill, 2007.

66 Tilman and Hill, 2007.

67 Hooijer and others, 2006.

68 A point I take from Monbiot, 2007.

69 The account that follows is particularly based on Monbiot, 2007; Food and Agriculture Organisation,

2006; Tilman and Hill, 2007; and Hooijer and others, 2006.

70 Malking, 2007.

71 The account that follows is heavily influenced by Romm, 2004.

72 German Aerospace Center (DLR), 2006, p13.

73 Romm, 2004, p76.

74 For the arguments against CCS see Rochon and others, 2008.

75 Rochon and others, 2008, p27.

76 The best book on Peak Oil is Simmons, 2005. Also very useful are Leggett, 2005; Deffeyes, 2001; and McKillop, ed., 2005. For natural gas "cliffs" see Darley, 2004.

77 Leggett, 2005, p45.

78 See Lima and others, 2007, and McCully, 2006a.

79 McCully, 2006b; World Commission on Dams, 2000; and Roy, 1999.

80 Christian Aid, 2007.

81 Detailed arguments about the dangers of nuclear power can be found in Empson, 2006, and Caldicott, 2006.

82 Greenpeace, 2006.

83 It is also worth remembering that when ozone (a form of oxygen) (O) breaks down methane (CH$_4$), the product is an equivalent amount of CO$_2$.

84 Stephens and others, 2007.

85 Hooijer and others, 2006, pp17-24.

86 Rodrigues, 2004.

Part Three

1 The discussion of profits that follows is based on Neale, 2004, pp7-23; Brenner, 2002 and 2006; and Harman, 1999a.

2 Brenner, 2002, p21. I have rounded up the numbers here. The statistics for different countries are not directly comparable, as they all use different accounting systems to calculate profits. The point is that in each country the falls were broadly similar.

3 Brenner, 2006, p7.

4 Brenner, 2002, p33.

5 The analysis of neoliberalism that follows is developed at much greater length in Neale, 2004. Other particularly useful books are Klein, 2007; Harvey, 2005; Whitfield, 2001; Jain, 2001; Bond, 2000; Green, 2003; and Allen, 2007.

6 For good studies of privatisation and subcontracting, see Monbiot, 2000;

Murray, 2001; Wolmar, 2005; Turshen, 1999; and Pollock, 2004.

7 For good studies of the process in health, see Pollock, 2004; Turshen, 1999; Abraham, 1993; and Neale, 1983.

8 See Pollock, 2004.

9 For more on prisons and racism in the US see Neale, 2004, pp87-111.

10 This is what led Tony Cliff, correctly in my opinion, to describe the Stalinist states as *bureaucratic state capitalist*. See Cliff, 1996.

11 I take this insight from Miller, 2005.

12 Wallace, 2005, pp188-189.

13 Klein, 2007, is very good on the links between defeating the workers movement and pushing through neoliberalism.

14 Nordlund, 1998.

15 Van Wersch, 1992.

16 Callinicos and Simons, 1985.

17 For why the Sandanistas lost see Gonzales, 1990, and Lancaster, 1994.

18 For overproduction see David Harvey, 1982. For global competition see Brenner, 2002 and 2006. For the classic Marxist theory of the tendency of the rate of profit to fall, see Harman, 1999a, and Marx, 1981.

19 Neale, 2004, pp73-75 and 113-15; Mishel, Bernstein and Bushey, 2003.

20 See Brenner, 2006, p7; Neale, 2004, pp19-20; and Harman, 2007.

21 See Introduction, note 1.

22 Table from Bradford, 2006, p105. I have rounded the figures up to the nearest full number.

23 Cameron and de Vries, 2006. The details on particular companies that follow are also taken from this article.

24 This section relies on Bradsher, 2003, ever such a fun book. Doyle, 2000, is also useful.

25 The figures are somewhat misleading to British readers. The US gallon is four fifths of the British "imperial" gallon. American government mileage tests are also more forgiving than in other countries, because they are done in the lab, rather than in real road conditions.

26 Bradsher, 2003, p96.

27 Bradsher, 2003, pp101-102.

28 Bradsher, 2003, pp95-97

29 Bradsher, 2003, pp76-80 and 264-265.

30 My account of the history of the oil industry relies particularly on McQuaig, 2004, and also on Shah, 2006, and Yeomans, 2005.
31 McQuaig, 2004, pp208-209.
32 McQuaig, 2004, p215.
33 Strictly speaking about two and a half times. But the Chinese figures for foreign investment are vastly inflated by Chinese money moving out and then being reinvested immediately, because companies with foreign investment get preferential treatment in China.

Part Four
1 See all the reports of the Intergovernmental Panel on Climate Change. They are published in book form by Cambridge University Press and are all listed in the bibliography at the end of the book. They are also available online at www.ipcc.ch/ipccreports/index.htm.
2 I take the phrase "carbon club" from Legget, 1999.
3 For the carbon corporations, denial and the press, see Gelbspan, 1998; Gelbspan, 2004, pp37-86; and Monbiot, 2006, pp20-42.
4 Gelbspan, 2004, p53.
5 Leggett, 1999.
6 See Wyler, 2004, for a detailed and decent account of the tensions between money, politics and vision.
7 Foster, 2002, has a chapter that gives a sensitive and complex explanation of the political conflicts between environmentalists and unions over old forests in the Pacific Northwest. Wohlforth, 2004, gives a nuanced description of a similar sort of contradictory and changing consciousness among Eskimo in Alaska.
8 For a particularly eloquent, funny and angry account of this sort of thing, see Geoghegan, 1991.
9 See, for instance, Lovelock, 2006.
10 For good accounts of the anti-capitalist movement see Thomas, 2000, on Seattle; Desai, 2002, on Durban; and Neale, 2002, on Genoa.
11 For the anti-war movement start with Murray and German, 2005.
12 Bows and others, 2006, pp163-166.

13 "Big Business urges G8 global warming action", *Financial Times*, 10 June 2005.
14 This and the next two paragraphs are based on Woodward, 1994, pp91-93 and 141-143.
15 Kennedy, 2004.
16 I have taken the text of Kennedy's speech from an email that circulated widely on the day.
17 For more on the global climate demonstrations, and if you want to get involved, go to www.globalclimate campaign.org. The account that follows is based on my own experience organising demonstrations, and is not the official view of the Global Climate Campaign, which unites people with a very wide range of views.
18 For more on Step It Up, go to McKibben, 2007.
19 Rogers, 2005, especially pp129-182. A shorter version of the argument can be found in Rogers, 2006.
20 Rogers, 2005, pp141-142.
21 Rogers, 2005, p144.
22 Rogers, 2005, pp144-145.
23 Rogers, 2005, p4.
24 The technical phrase is that money is "fungible".
25 Lohman, 2006, is the best explanation of what's wrong with cap and trade schemes.
26 Lohman, 2006, pp108-109.
27 Lohman, 2006, pp108-109.
28 CDMs are tightly regulated by the UN. No tree planting is allowed, because the scams are so easy. CDMs were originally inserted into Kyoto to tempt poor countries by promising them help from rich countries with renewable energy. The idea is that carbon offsets from CDMs will allow companies or governments in rich countries to make up for their failures to cut emissions as promised in the Kyoto agreement.
 In practice, as of August 2006, 2 percent of the total permits in CDMs were for building renewable energy. Of the 265 CDM projects in the poor countries generating credits, just seven accounted for almost three quarters of the total. All seven of these projects were at factories in India and China

making air conditioners, fridges and nylon. These factories all produced two very rare and very powerful greenhouse gases, HFC-23 and N20. HFC-23, for instance, has 11,700 times the warming effect of CO_2.

All the seven companies had to do was bolt on some extra machinery to catch the escaping gases. They then had very large carbon credits to sell through the UN and the Kyoto Protocol, and it cost them hardly anything. Of course the factories ought to do that. But no factory on earth should be permitted to emit such gases. It would be no problem for the Indian and Chinese governments simply to forbid such emissions. If the governments in the rich countries simply refused to import goods made in such factories, the emissions would also stop. See Lohman, 2006, p164.

29 Good accounts of the Contraction and Convergence approach can be found in Mayer, 2000; Hillman, 2004; and from a more left-wing perspective, Sims, 2005.

Part Five

1 I have found Klinenberg, 2002, and Klein, 2007, very useful for understanding the relationships between capitalism, neoliberalism and climate disasters.

2 The account of Katrina that follows relies on Horne, 2006; Brinkley, 2006; Tidwell, 2003 and 2006; Reed, 2006; McQuaid and Schleifstein, 2006; Dyson, 2006; Troutt, 2006; van Heerden and Bryan, 2006; and Spike Lee's documentary *When the Levees Broke* (BBC4, 2006).

3 For the changing scientific understanding of the relationship between hurricanes and climate change see Mooney, 2007, and McQuaid and Schleifstein, 2006, pp345-356. For the evidence for increase intensity of tropical storms, see Emmanuel, 2005 and Webster and others, 2005.

4 And as the air warms, the air pressure inside the eye falls. That fall literally pulls the surface of the ocean up a few feet, and increases the height of the surge.

5 McQuaid and Schleifstein, 2006, p350.

6 McQuaid and Schleifstein, 2006, p85.

7 Tidwell, 2006, p25.

8 Dyson, 2006, pp72 and 82-83, and Tidwell, 2003. In fact John McQuaid and Mark Schleifstein of the *New Orleans Times-Picayune* won a Pulitzer Prize, the top award in US journalism, for their "Washing Away" series in 2002.

9 Dyson, 2006, p78.

10 Dyson, 2006, pp58-59.

11 Brinkley, 2006, pp1-3.

12 See Klinenberg, 2002, for Chicago, and Klein, 2007, for Iraq.

13 Dyson, 2006, pp118 and 120-121.

14 Mooney, 2007, pp9-10.

15 Rose, 2007, gives a detailed and moving picture of what life was like in New Orleans in the first two years after the hurricane.

16 Adolf Reed, Jnr (2006a and 2006b) has made this argument well. One of the many strengths of Spike Lee's *When the Levees Broke* is the way he shows quietly how both black and white working people had their lives devastated.

17 Adolf Reed, Jnr, in Troutt, 2006, p30.

18 The following account of the Sudanese national economy from 1955 to 1990 relies particularly on Brown, 1988 and 1992; Hussein, 1988; and Karim, 1988. Alier, 1992, is essential for the roots of the civil war, especially his chapter on oil, pp236-245.

19 Brown, 1992, p211. Brown rightly argues that the IMF was less harsh on Sudan before 1984 because the government in Khartoum was then allied with the USA. But this shifted after 1984, and in any case less harsh was still harsh—see Hussein, 1988, and Karim, 1988.

20 For the science of this see Zeng, 2003; Gianni, Sanavan and Chung, 2003; and Flannery, 2005, pp124-7.

21 For the complexity of traditional politics and ethnicity in Darfur, see Daley, 2007; De Waal 2005; Holy, 1980; Barth, 1967; and Haaland, 1969.

22 This account of the famine owes much to De Waal, 2005. For the politics that

surrounded the famine, and the years 1984-89, I have also relied on Daley, 2007; Prunier, 2005 and 2007; and Keen, 1994.

23 Prunier, 2005, pp54-56.

24 This understanding of the famine is based on De Waal, 2005.

25 For just what it means to a herder to lose his animals, see the account of the plight of poor herders and their oppression by flock owners in neighbouring Kordofan in the 1960s in Asad, 1970.

26 The following discussion of farmers and herders in Darfur relies particularly on Daley, 2007; De Waal, 2005; Keen, 1994; and Holy, 1980. I have also relied on three good ethnographies of neighbouring ethnic groups: Cunnison, 1966, on Baggara Arabs in Kordofan; Asad, 1970, on camel herding Arabs in Kordofan; and Tubiana and Tubiana, 1977, on the Zaghawa in Chad. It also reflects wide reading on pastoral nomads, my own research with nomads in Afghanistan in 1971-73, and two months travelling in Kordofan in 1968. Much of the literature on pastoral nomads in the Middle East is misleadingly rosy on class relations among nomads. The exception for the Sudan is Asad, 1970, for Iran, Black-Michaud, 1986, and Bradburd, 1980; and for Afghanistan, Tapper, 1991. The literature on northern Sudanese pastoralists suggests that they are not that different. The herders of the southern Sudan were running rather different systems, although the ethnographers may have been exaggerating the equality in those societies, and Hutchinson, 1996, describes a widespread cash market for cattle among the the Nuer by the 1980s.

27 This account of the cleansing of the oil fields relies on Keen, 1994.

28 Keen, 1994, pp99-100.

29 For the war in Chad see Nolutshungu, 1996; and Burr and Collins, 1994.

30 I take the phrase "globalisation of ethnic cleansing" from Prunier, 2005.

31 My understanding of the current war relies particularly on Prunier, 2005 and 2007; Daley, 2007; Flint and De Waal, 2005; Haggar, 2007; Marchal, 2007; Joseph Tubiana, 2007; and Fadul and Tanner, 2007.

32 For the sake of simplicity, I will leave out the role of the governments of Ethiopia, Eritrea, Britain, Uganda, Nigeria, Israel and the Central African Republic, all of whom were also involved.

33 Part of it certainly came from Darfuri migrants working across the Middle East. Some of it was coming from Chad. There was also at least political encouragement from France, Israel, and from Christian evangelicals.

34 Prunier, 2005, pp95-96.

35 In fact, what was happening in Darfur was not like the Nazi genocide that tried to exterminate all Jews. It was more like American bombing that killed more than a million Vietnamese villagers in attempt to drive the Viet Cong guerrillas off the land.

36 De Waal, 2005, pp78-104.

37 De Waal, 2005, p80.

38 Lee, 1979.

39 For good explanations of the Russian Revolution and its subsequent degeneration, see Haynes, 2002, and Rees, 1997. Both strongly challenge the idea that Lenin laid the foundations for Stalin's dictatorship.

40 For the argument about equality and sharing in traditional hunting and gathering societies, see Lee, 1979, Sahlins, 1972, and Leacock, 1981. These arguments have been criticised on two grounds. One is that they are largely based on romanticising groups of hunters in the 20th century who were in fact deeply enmeshed in the global and local economy, and were also the most powerless and oppressed groups in the regions they lived in (see, for instance, Gordon, 1991). The other is that there is evidence for warfare and murder in traditional gathering societies. Both of these criticisms are justified. But in my view they do not contradict the central point of Lee, Sahlins, and Leacock's argument about the circumstances of pre-agricultural society. For the development of human history from the invention of agriculture on, the best place to start is Harman, 1999b.

BIBLIOGRAPHY

NOTE: Website references are generally just to the site address rather than the specific page as websites are constantly being changed.

Abraham, Laura Kaye, *Mama Might Be Better off Dead: the Failure of Urban Health Care in America* (University of Chicago Press, Chicago, 1993).

Alier, Abel, *Southern Sudan: Too Many Agreements Dishonoured* (Ithaca Press, Reading, 1992).

Allen, Kieran, *The Corporate Takeover of Ireland* (Irish Academic Press, Dublin, 2007).

Alley, Richard B, *The Two-Mile Time Machine: Ice Cores, Abrupt Climate Change and our Future* (Princeton University Press, Princeton, 2000).

Archer, Christina L, and Jacobsen, Mark Z, "Evaluation of Global Wind Power", *Journal of Geophysical Research— Atmospheres*, vol 110 (2005).

Asad, Talal, *The Kababish Arabs* (Hurst, London, 1970).

Barnett, Tony, and Abdelkarim, Abbas (eds), *Sudan: State, Capital and Transformation* (Croom Helm, London, 1988).

Barrett, Mark, *A Renewable Electricity System for the UK* (University College London, 2006), available at www.cbes.ucl.ac.uk

Barry, John, *Rising Tide: The Great Mississippi Flood of 1927 and How it Changed America* (Simon and Schuster, New York, 1997).

Barth, Fredrik, "Economic Spheres in Darfur", in Raymond Firth, ed, *Themes in Economic Anthropology* (Tavistock, London, 1967).

Beatty, Jack, *Age of Betrayal: The Triumph of Money in America, 1865-1900* (Alfred A. Knopf, New York, 2007).

Benton, Michael J, *When Life Nearly Died: The Greatest Mass Extinction of All Time* (Thames and Hudson, London, 2003).

"Big Business urges G8 global warming action", *Financial Times*, 10 June 2005.

Black-Michaud, Jacob, *Sheep and Land* (Cambridge University Press, Cambridge, 1986).

Bond, Patrick, *Against Global Apartheid: South Africa Meets the World Bank, IMF and International Finance* (University of Cape Town Press, Cape Town, 2000).

Bows, Alice, Mander, Sarah, Starkey, Richard, Bleda, Mercedes, and Anderson, Kevin, *Living within a Carbon Budget* (Tyndall Centre, Manchester, 2006).

Bradburd, Daniel, "Never Give a Shepherd an Even Break: Class and Labour among the Komachi", *American Ethnologist*, 7:603-620 (1980).

Bradford, Travis, *Solar Revolution* (MIT Press, Cambridge MA, 2006).

Bradsher, Keith, *High and Mighty: The Dangerous Rise of the SUV* (second edition, Public Affairs, New York, 2003).

Breman, Jan, *Footloose Labour: Working in India's Informal Economy* (Cambridge University Press, Cambridge, 1996).

Breman, Jan, *The Labouring Poor in India* (Oxford University Press, Delhi, 2003).

Brenner, Robert, *The Boom and the Bubble: The USA in the World Economy* (second edition, Verso, London, 2002).

Brenner, Robert, *The Economics of Global Turbulence* (Verso, London, 2006).

Brinkley, Douglas, *The Great Deluge: Hurricane Katrina, New Orleans, and the Mississippi Gulf Coast* (William Morrow, New York, 2006).

Brown, Richard, "A Background Note on the Final Round of Economic Austerity Measures Imposed by the Numeiry Regime: June 1984-March 1985", in Barnett and Abdulkarim, 1988.

Brown, Richard, *Public Debt and Private Wealth: Debt, Capital Flight and the IMF in Sudan* (Macmillan, London, 1992).

Burnett, John, *Plenty and Want: A Social History of Diet in England from 1815 to the Present Day* (Scolar, London, 1979).

Burr, J, and Collins, R O, *Africa's Thirty Years War: Libya, Chad, and the Sudan, 1963-1993* (Westview, Boulder, 1994).

Caldicott, Helen, *Nuclear Power is Not the Answer* (The New Press, New York, 2006).

Callinicos, Alex, and Simons, Mike, *The Great Strike: The Miners' Strike 1984-5 and its Lessons* (Bookmarks, London, 1985).

Cameron, Alisdair and De Vries, Eize, "Top of the List", *Renewable Energy World*, February 2006.

Caulfield, Catherine, *Masters of Illusion: The World Bank and the Poverty of Nations* (Macmillan, London, 1997).

Christian Aid, *Human Tide: The Real Migration Crisis* (2007), available at www.christianaid.org.uk.

Clark, Nicola, "Boeing Scores with Dreamliner Order", *New York Times*, 20 June 2007.

Cliff, Tony, *State Capitalism in Russia* (Bookmarks, London, 1996 [1948]).

Cox, John D, *Abrupt Climate Change and What It Means for Our Future* (Joseph Henry Press, Washington DC, 2005).

Cunnison, Ian, *Baggara Arabs: Power and Lineage in a Sudanese Tribe* (Clarendon Press, Oxford, 1966).

Daley, M W, *Darfur's Sorrow* (Cambridge University Press, Cambridge, 2007).

Darley, Julian, *High Noon for Natural Gas* (Chelsea Green, White River Junction, 2004).

Deffeyes, Kenneth, *Hubbert's Peak: The Impending World Oil Shortage* (Princeton University Press, Princeton, 2001).

Desai, Ashwin, *We are the Poors: Community Struggles in Post-Apartheid South Africa* (Monthly Review Press, New York, 2002).

Doyle, Jack, *Taken for a Ride: Detroit's Big Three and the Politics of Pollution* (Four Walls Eight Windows, New York, 2000).

Dyson, Michael Eric, *Come Hell or High Water: Hurricane Katrina and the Color of Disaster* (Basic Books, New York, 2006).

Emmanuel, Kerry, "Increasing Destructiveness of Tropical Cyclones over the Past Thirty Years", *Nature* 436 (2005), pp686-88.

Empson, Martin, *Climate Change: Why Nuclear Power is Not the Answer* (Socialist Worker, London, 2006).

Fadul, Abdul Jaffar, and Tanner, Victor, "Darfur after Abuja, A View from the Ground", in De Waal, 2007, pp284-313.

Flannery, Tim, *The Weather Makers* (London, 2005).

Flint, Julie, and De Waal, Alex, *Darfur: A Short History of a Long War* (Zed, London, 2005).

Food and Agriculture Oranization, "Overview" and "Coarse Grains" in *Food Outlook* No 2, December 2006, at www.fao.org.

Foster, John Bellamy, *Marx's Ecology: Materialism and Nature* (Monthly Review Press, New York, 2000).

Foster, John Bellamy, *Ecology against Capitalism* (Monthly Review Press, New York, 2002).

Gelbspan, Ross, *The Heat is On* (Perseus, Cambridge MA, 1998).

Gelbspan, Ross, *Boiling Point* (Basic Books, New York, 2004).

Geoghegan, Thomas, *Which Side Are You On? Trying to Be for Labor When It's Flat on its Back* (Farrar Strauss and Giroux, New York, 1991).

German Aerospace Center (DLR), *Concentrating Solar Power for the Mediterranean Region* (2005), available at www.dlr.de/tt/trans-med or at www.trec-uk.org.uk.

German Aerospace Center (DLR), *Trans-Mediterranean Interconnection for Concentrating Solar Power for the Mediterranean Region* (2006), available at www.dlr.de/tt/med-csp or at www.trec-uk.org.uk.

Gianni, A, Sanavan, R, and Chung, P, "Oceanic Forcing of Sahel Rainfall on Interannual to Interdecadal Timescales", *Science*, 2003, 302: 1027-1030.

Gilbert, Simon, "China's Strike Wave", *International Socialism*, 107 (2005).

Gipe, Paul, *Wind Power* (second edition, Chelsea Green, White River Junction, 2004).

Gipe, Paul, "Design as if People Matter: Aesthetic Guidelines for a Wind Power Future", in Pasqualetti, Gipe and Richter, 2002, pp173-212.

Gonzales, Mike, *Nicaragua: What Went Wrong?* (Bookmarks, London, 1990).

Gordon, Robert, *The Bushman Myth:*

Stop Global Warming: Change the World

Making of a Namibian Underclass (Westview, Boulder, 1991).

Green, Duncan, Silent Revolution: The Rise and Crisis of Market Economies in Latin America (Monthly Review Press, New York, 2003).

Greenhalgh, Susan, and Winkler, Edwin, Governing China's Population: from Leninist to Neoliberal Biopolitics (Stanford University Press, Stanford, 2005).

Greenpeace, The Chernobyl Catastrophe: Consequences on Human Health (Greenpeace International, Amsterdam, 2006).

Greenwald, Robert, director, Wal-Mart— The High Cost of Prices (DVD, Tartan Video, 2006).

Haggar, Ali, "The Origins and Organization of the Janjawid in Darfur," in De Waal, 2007, pp113-139.

Haaland, Gunnar, "Economic Determinants of Ethnic Processes", in Barth, Fredrik, ed, Ethnic Groups and Boundaries (Allen Unwin, London, 1969).

Hansen, James, "Why We Can't Wait", The Nation, 7 May 2007.

Hansen, J, Sato, Mki, Kharocha, P, Russell, G, Lea, D W, and Siddall, M, "Climate Change and Trace Gases", Philosophical Transactions of the Royal Society A, 365:1925-54 (2007).

Harman, Chris, Explaining the Crisis: A Marxist Reappraisal (Bookmarks, London, 1999a).

Harman, Chris, A People's History of the World (Bookmarks, London, 1999b).

Harman, Chris, "Snapshots of Capitalism Today and Tomorrow", International Socialism 113 (2007).

Harvey, David, The Limits to Capital (Basil Blackwell, Oxford, 1982).

Harvey, David, A Brief History of Neoliberalism (Oxford University Press, Oxford, 2005).

Hawken, Paul, Lovins, Amory, and Lovins, L Hunter, Natural Capitalism: Creating the Next Industrial Revolution (Back Bay Books, New York, 2000).

Haynes, Mike, Russia: Class and Power 1917-2000 (Bookmarks, London, 2002)

Heerden, Ivor Van, and Bryan, Mike, The Storm: What Went Wrong and Why

During Hurricane Katrina—The Inside Story from One Louisiana Scientist (New York, 2006).

Hickman, Martin, "Only Wealthiest will be able to Afford Solar Panels", Independent, 10 May 2007.

Hillman, Mayer with Fawcett, Tina, How We Can Save the Planet (Penguin, London, 2004).

Holy, Ladislav, Neighbours and Kinsmen: A Study of the Berti People of Darfur (Hurst, London, 1974).

Holy, Ladislav, "Drought and Change in a Tribal Economy: the Berti of Northern Darfur", Disasters, 4:65-72 (1980).

Hooijer, A, Silvius, M, Woosten, H, Page, S, PEAT—CO_2, Assessment of CO_2 Emissions from Drained Peatlands in SE Asia (Dreft Hyraulics report Q343, 2006), available from www.wetlands.org.

Horne, Jed, Breach of Faith: Hurricane Katrina and the Near Death of a Great American City (Random House, New York, 2006).

Hussein, Mohammed Nureldin, "The IMF and Sudanese Economic Policy", in Barnett and Abdulkarim, 1988, pp55-72.

Hutchinson, Sharon, Nuer Dilemmas: Coping with Money, War, and the State (Uiversity of California Press, Berkeley, 1996).

Intergovernmental Panel on Climate Change, Scientific Assessment of Climate Change (Cambridge University Press, Cambridge, 1990).

Intergovernmental Panel on Climate Change, Impacts Assessment of Climate Change (Cambridge University Press, Cambridge, 1990).

Intergovernmental Panel on Climate Change, The IPCC Response Strategies (Cambridge University Press, Cambridge, 1990).

Intergovernmental Panel on Climate Change, Climate Change 1995: The Science of Climate Change (Cambridge University Press, Cambridge, 1995).

Intergovernmental Panel on Climate Change, Climate Change 1995: Impacts, Adaptation and Mitigation of Climate Change (Cambridge University Press, Cambridge, 1995).

Intergovernmental Panel on Climate Change, *Climate Change 1995: Economic and Social Dimensions of Climate Change* (Cambridge University Press, Cambridge, 1995).

Intergovernmental Panel on Climate Change, *Climate Change 2001: The Scientific Basis* (Cambridge University Press, Cambridge, 2001).

Intergovernmental Panel on Climate Change, *Climate Change 2001: Impacts, Adaptation and Vulnerability* (Cambridge University Press, Cambridge, 2001).

Intergovernmental Panel on Climate Change, *Climate Change 2001: Mitigation* (Cambridge University Press, Cambridge, 2001).

Intergovernmental Panel on Climate Change, *Climate Change 2007: The Physical Science Basis* (Cambridge University Press, Cambridge, 2007).

Intergovernmental Panel on Climate Change, *Climate Change 2007: Impacts, Adaptation and Vulnerability* (Cambridge University Press, Cambridge, 2007).

Intergovernmental Panel on Climate Change, *Climate Change 2007: Mitigation of Climate Change* (Cambridge University Press, Cambridge, 2007).

Jain, Neerai, *Globalisation or Recolonisation* (Alaka Joshi, Pune, 2001).

Karim, Hassan Gad, "Sudanese Government Attitudes Towards Foreign Investment—Theory and Practice", in Barnett and Abdulkarim, 1988, pp37-44.

Keen, David, *The Benefits of Famine: A Political Economy of Famine and Relief in Southwestern Sudan, 1983-1989* (Princeton University Press, Princeton, 1994).

Kennedy, Robert F Jr, *Crimes Against Nature; How George W Bush and His Corporate Pals Are Plundering the Country and Hijacking Our Democracy* (HarperCollins, New York, 2004).

Klein, Naomi, *The Shock Doctrine: The Rise of Disaster Capitalism* (Allen Lane, London, 2007).

Klinenberg, Eric, *Heat Wave* (Chicago University Press, Chicago, 2002).

Koistinen, Paul, *Arsenal of World War II: The Political Economy of American Warfare* (University Press of Kansas, Lawrence, 2004).

Krause, Elizabeth, *Dangerous Demographies: The Scientific Manufacture of Fear* (Corner House, Sturminster Newton, Dorset, 2006).

Lancaster, Roger, *Life is Hard: Machismo, Danger and the Intimacy of Power in Nicaragua* (University of California Press, Berkeley, 1994).

Lankton, Larry, "Autos to Armaments: Detroit becomes the Arsenal of Democracy", *Michigan History* (November/December 1991), pp42-49.

Leacock, Eleanor, *Myths of Male Dominance* (Monthly Review Press, New York, 1981).

Lee, Richard, *The !Kung San: Men, Women and Work in a Foraging Society* (Cambridge University Press, Cambridge, 1979).

Lee, Spike, director, *When the Levees Broke* (BBC4, 2006).

Leggget, Jeremy, *The Carbon War: Global Warming and the End of the Oil Era* (Penguin, London, 1999).

Leggett, Jeremy, *Half Gone: Oil, Gas, Hot Air and the Global Energy Crisis* (Portobello, London, 2005).

Lima, Ian, Ramos, Fernando, Bambace, Luis, and Rosa, Reinaldo, "Methane Emissions from Large Dams as Renewable Energy Resources: A Developing Nations Perspective", *Mitigation and Adaptation Strategies in Global Change* (2007).

Lohman, Larry, 2006, *Carbon Trading: A Critical Conversation on Climate Change, Privatisation and Power* (Special Issue of *Development Dialogue*, no 48, Stockholm, 2006).

Lovelock, James, *The Revenge of Gaia: Earth's Climate Crisis and the Fate of Humanity* (Basic Books, New York, 2006).

Lovins, Amory, Datta, E Kyle, Bustnes, Odd-Even, and Koomy, Jonathan, *Winning the Oil Endgame: Innovation for Profit, Jobs and Security* (Rocky Mountain Institute, Snowmass CO, 2004).

Stop Global Warming: Change the World

McCully, Patrick, *Fizzy Science: Loosening the Hydro Industry's Grip on Reservoir Greenhouse Gas Emissions Research* (International Rivers Network, Berkley, 2006a), available at www.irn.org.

McCully, Patrick, *Silenced Rivers: The Ecology and Politics of Large Dams* (Zed, London, 2006b).

McGregor, Richard, "Beijing Censored Pollution Report", *Financial Times*, 3 July 2007.

McKibben, Bill, and the Step-It-Up Team, *Fight Global Warming Now: The Handbook for Taking Action in your Community* (Holt, New York, 2007).

McKillop, Andrew (ed), *The Final Energy Crisis* (Pluto, London, 2005).

McQuaid, John, and Schleifstein, Mark, *Path of Destruction: The Devastation of New Orleans and the Coming Age of Superstorms* (Little Brown, New York, 2006);

McQuaig, Linda, *It's the Crude, Dude: War, Big Oil and the Fight for the Planet* (Doubleday Canada, Toronto, 2004).

Malking, Elisabeth, "Thousands in Mexico City Protest Rising Food Prices", *New York Times*, 1 February 2007.

Mamdani, Mahmood, *The Myth of Population Control* (Monthly Review Press, New York, 1975).

Marchal, Roland, "The Unseen Regional Implications of the Crisis in Darfur", in De Waal, 2007, pp173-198.

Mayer, Aubrey, *Contraction and Convergence: The Global Solution to Climate Change* (Green Books, Dartington, 2000).

Mayewski, Paul, and White, Frank, *The Ice Chronicles* (University Press of New England, Hanover, 2002).

Marx, Karl, *Capital: Volume Three* (Penguin, London, 1981).

Mayer, Aubrey, *Contraction and Convergence: The Global Solution to Climate Change* (Green Books, Dartington, 2000).

Mieszkowicz, Joanna, "Methodology of Carbon Dioxide Emission from Transport", Aeris Futoro Foundation, available at www.aeris.eko.org.p/ang' kaljulator/methodology_transport.pdf

Miller, David, "Propaganda-Managed Democracy: the UK and the Lessons of Iraq", in Panitch and Leys, *Telling the Truth: Socialist Register* 2006 (Merlin, London, 2005), pp134-145.

Mishel, Lawrence, Bernstein, Jared, and Bushey, Heather, *The State of Working America, 2002/2003* (Cornell University Press, Ithica, 2003).

Monbiot, George, *Captive State: The Corporate Takeover of Britain* (Macmillan, London, 2000).

Monbiot, George, *Heat: How to Stop the Planet Burning* (Allen Lane, London, 2006).

Monbiot, George, "A Lethal Solution" posted on 27 March 2007 at www.monbiot.com.

Mooney, Chris, *Storm World: Hurricanes, Politics and the Battle over Global Warming* (Harcourt, London, 2007).

Murray, Andrew, *Off the Rails: The Crisis on Britain's Railways* (Verso, London, 2001).

Murray, Andrew, and German, Lindsey, *Stop the War: the Story of Britain's Biggest Mass Movement* (Bookmarks, London, 2005).

Navarro, Vincent, "The Worldwide Class Struggle", *Monthly Review*, 58.4, 2006.

Neale, Jonathan, *Memoirs of a Callous Picket: Working for the NHS* (Pluto, London, 1983).

Neale, Jonathan, *You are G8, We are 6 Billion* (Vision, London, 2002).

Neale, Jonathan, *What's Wrong with America? How the Rich and Powerful Have Changed America and Now Want to Change the World* (Vision, London, 2004).

Ngai, Pun, *Made in China: Women Factory Workers in a Global Marketplace* (Duke University Press, Durham, 2005).

Nielsen, Frode, "A Formula for Success in Denmark", in Pasqualetti, Gipe and Richter, 2002, pp115-132.

Nolutshungu, Sam, *Limits of Anarchy: Intervention and State Formation in Chad* (University Press of Virginia, Charlottesville, 1996).

Nordlund, Willis, *Silent Skies: the Air Traffic Controllers' Strike* (Praeger, Westport, 1998).

Paine, Chris, director, *Who Killed the Electric Car?* (DVD, Sony, 2006).

Panitch, Leo, and Colin Leys, Colin (eds), *Socialist Register 2007: Coming to*

Terms with Nature (Merlin, London, 2006).

Pasqualetti, Martin, "Living with Wind Power in a Hostile Landscape", in Pasqualetti, Gipe and Richter, 2002, pp153-172.

Pasqualetti, Martin, Gipe, Paul, and Richter, Robert (eds), Wind Power in View: Energy Landscapes in a Crowded World (Academic Press, San Diego, 2002).

Pearce, Fred, The Last Generation (Eden Project Books, London, 2006). Also published as With Speed and Violence (Beacon Press, Boston, 2007).

Penner, J, Lister, D, Griggs, Dekker, D, and McFarland, M, Aviation and the Global Atmosphere (Cambridge University Press, Cambridge, 1999).

Perry, Elizabeth, Shanghai on Strike (Stanford University Press, Palo Alto, 1995).

Pollock, Allyson, NHS Plc: The Privatisation of our Health Care (Verso, London, 2004).

Population Division of the Department of Economic and Social Affairs of the United Nations Secretariat, World Population Prospects: the 2006 Revisions, accessed at http://esa.un.org/unpp.

Prunier, Gerard, Darfur: The Ambiguous Genocide (Hurst, London, 2005).

Prunier, Gerard, Darfur: The Ambiguous Genocide (second edition, Cornell University Press, Ithica, 2007).

Rabe, Barry, Statehouse and Greenhouse: The Stealth Politics of American Climate Change Policy (Brookings, Washington, 2003).

Rao, Mohan, From Population Control to Reproductive Health: Malthusian Arithmetic (Sage, London, 2004).

Reed, Adolf Jr, "The Real Divide", in Troutt, 2006, pp63-70.

Reed, Adolf Jr, "Class-ifying the Hurricane", in Betsy Reed, 2006b, pp27-32.

Reed, Betsy (ed), Unnatural Disaster: The Nation on Hurricane Katrina (Nation Books, New York, 2006).

Rees, John, In Defence of October (Bookmarks, London, 1997).

Roaf, Sue, Crichton, David, and Nicol, Fergus, Adapting Buildings and Cities for Climate Change: A 21st Century Survival Guide (Architectural Press, Oxford, 2005).

Rochon, Emily and others, False Hope: Why Carbon Capture and Storage Won't Save the Climate (Greenpeace International, Amsterdam, 2008).

Rodrigues, Moog, Gaudalupe, Maria, Global Environmentalism and Local Politics: Transnational Advocacy Networks in Brazil, Ecuador and India (State University of New York Press, Albany, 2004).

Rogers, Heather, Gone Tomorrow: The Hidden Life of Garbage (The New Press, New York, 2005).

Rogers, Heather, "Garbage Capitalism's Green Commerce", in Panitch and Leys, 2006.

Romm, Joseph, The Hype about Hydrogen: Fact and Fiction in the Race to Save the Climate (Island Press, Washington, 2004).

Rose, Chris, 1 Dead in Attic (Simon and Schuster, New York, 2007).

Roy, Arundathi, The Cost of Living (Flamingo, London, 1999).

Sahlins, Marshall, Stone Age Economics (Aldine-Atherton, Chicago, 1972).

Scheer, Hermann, The Solar Economy (Earthscan, London, 2002).

Scheer, Hermann, A Solar Manifesto (second edition, James and James, London, 2005).

Shah, Sonia, Crude: The Story of Oil (Seven Stories, New York, 2006).

Simmons, Matthew, Twilight in the Desert: The Coming Saudi Oil Shock and the World Economy (Wiley, Hoboken, 2005)

Sims, Andrew, Ecological Debt: The Health of the Planet and the Wealth of Nations (Pluto, London, 2005).

Sperling, Daniel, and Salon, Deborah, Transportation in Developing Countries: An Overview of Greenhouse Gas Reduction Strategies (Pew Center on Global Climate Change, 2002), access at www.pewclimate.org.

Stephens, Britton and others, "Weak Northern and Strong Tropical Land Carbon Uptake from Vertical Profiles of Atmospheric CO_2", Science, 316:1732-35 (2007).

Stern, Nicholas, The Economics of

Stop Global Warming: Change the World

Climate Change: The Stern Review (Cambridge University Press, Cambridge, 2007).

Tapper, Nancy, Bartered Brides: Marriage and Politics in Northern Afghanistan (Cambridge University Press, Cambridge, 1991).

Thomas, Janet, The Battle in Seattle (Fulcrum, Golden CO, 2000).

Tidwell, Mike, Bayou Farewell: The Rich Life and Tragic death of Louisiana's Cajun Coast (Vintage, New York, 2003).

Tidwell, Mike, The Ravaging Tide: Strange Weather, Future Katrinas, and the Coming Death of America's Coastal Cities (Free Press, New York, 2006).

Tilman, David, and Hill, Jason, "Corn Can't Solve Our Problem", Washington Post, 25 March 2007.

Timmons Roberts, J, and Parks, Bradley C, A Climate of Injustice: Global Inequality, North-South Politics, and Climate Poverty (MIT Press, Cambridge, 2007).

Troutt, David Dante (ed), After the Storm: Black Intellectuals Explore the Meaning of Katrina (The New Press, New York, 2006).

Tubiana, M J, and Tubiana, Jerome, The Zaghawa from an Ecological Perspective (Balkena, Rotterdam, 1977).

Tubiana, Jerome, "Darfur: A War for Land?" in De Waal, 2007, pp68-91.

Turshen, Meredith, Privatizing Health Services in Africa (Rutgers University Press, New Brunswick, 1999).

Tyndall Centre, Decarbonising the UK: Energy for a Climate Conscious Future, available at www.tyndall.ac.uk.

De Waal, Alex, Famine that Kills: Darfur, Sudan (revised edition, Oxford University Press, Oxford, 2005).

De Waal, Alex (ed), War in Darfur and the Search for Peace (Global Equity Initiative, Harvard, Cambridge MA, 2007).

Wallace, David Foster, "Up Simba", in Consider the Lobster and Other Essays (Abacus, London, 2005), pp156-234.

Webster, Peter, Holland, Greg, Curry, Judith and Chung, Hai-Ru, "Changes in Tropical Cyclone Number, Duration and Intensity in a Warming Environment", Science, 309: 1844-1846 (2005).

Weizsacker, Ernst von, Lovins, Amory, and Lovins, L Hunter, Factor Four: Doubling Wealth, Halving Resource Use (Earthscan, London, 2001).

Wen, Dale, and Li, Minqi, "China: Hyper-Development and Environmental Crisis", in Panitch and Leys, 2006, pp130-146.

Wersch, Hubert van, Bombay Textile Strike, 1982-83 (Oxford University Press, Delhi, 1992).

White, Tyrene, China's Longest Campaign: Birth Planning in the People's Republic, 1949-2005 (Cornell University Press, Ithaca, 2006).

Whitfield, Dexter, Public Service or Corporate Welfare: Rethinking the Nation State in the Global Economy (Pluto, London, 2001).

Williams, Wendy, and Whitcomb, Robert, Cape Wind: Money, Celebrity, Class, Politics and the Battle for Our Energy Future on Nantucket Sound (Public Affairs, New York, 2007)

Wohlforth, Charles, The Whale and the Supercomputer: On the Northern Front of Climate Change (North Point Press, New York, 2004).

Wolmar, Christian, On the Wrong Line: How Ideology and Incompetence Wrecked Britain's Railways (Aurum, London, 2005).

Woodward, Bob, The Agenda: Inside the Clinton White House (Simon and Schuster, New York, 1994).

World Commission on Dams, Dams and Development (Earthscan, London, 2000).

Wyler, Rex, Greenpeace (Rodale, Emmaus PA, 2004).

Yeomans, Mathew, Oil: A Concise Guide to the Most Important Product on Earth (The New Press, New York, 2005).

Zeng, N, "Drought in the Sahel", Science, 2003, 303: 1124-27.

Zweiniger-Bargielowska, Ina, Austerity in Britain: Rationing, Controls and Consumption, 1939-1955 (Oxford University Press, Oxford, 2000).

INDEX

There is no entry under United States in this index as it can be found throughout the text.

Stop Global Warming: Change the World

Index

Acknowledgments

Two people have deeply shaped my understanding of environmental matters. This book would not have been possible without them. Linda Maher argued all through the 1980s about the possibility and necessity of combining environmentalism with socialism. Phil Thornhill's vision of a grassroots global climate movement has inspired me. I dedicate this book to them.

Nancy Lindisfarne and Ruard Absaroka encouraged me to write this book from the beginning. Nancy has edited and argued over every bit of every draft, and been my rock and my joy.

I am grateful to the people who have commented on various drafts and saved me from deep embarrassment: Jørn Andersen, Martin Empson, Chris Harman, Charlie Hore, Charlie Kimber, Fergus Nicol, Siobhan Neale, Ian Rappel, Peter Robinson, Jim Ruschill and John Sinha. And I am particularly grateful to Chris Nineham and Mark Thomas for wise, careful and persistent editing. Their work has made it a much better book.

I was steadily conscious in writing this book of the influence of Terry Neale and Marge Ruschill, and how much I would have liked to show it to them. And my thanks to all the climate activists in many countries who have taught me in the struggle.

Jonathan Neale
findjonathan@hotmail.com